METRIC MANUAL

METRIC MANUAL

By

Lawrence D. Pedde
Warren E. Foote
LeRoy F. Scott
Danny L. King
Dave L. McGalliard

GRAND RIVER BOOKS
1249 Washington Blvd. • Detroit, Michigan 48226 • 1980

Bibliographic note

This volume was originally published in 1978 by the U.S. Department of Interior, Bureau of Reclamation (GPO S/N 024-003-00129-5). The compilers were Lawrence D. Pedde, Warren E. Foote, LeRoy F. Scott, Danny L. King, and Dave L. McGalliard.

Library of Congress Cataloging in Publication Data

United States. Bureau of Reclamation.
   Metric manual.

   Bibliography: p. 251
   Includes index.
   1. Metric system--United States--Handbooks, manuals, etc. I. Pedde, Lawrence D.
QC92.U54U43   1980   389'.16'0973   79-29660
ISBN 0-8103-1020-1

PREFACE

By an act passed on July 28, 1866, the Congress authorized the metric system as a legal system of measurement in the United States. The United States was also one of the original signers of the Treaty of the Metre, which established an international metric system in 1875. This support for the metric system, however, did not include declaring that the metric system was this country's basic system of measurements. The English units of measurement were permitted to continue in use, and their existence was recognized indirectly through legislative provisions using the conventional measurements.

On December 23, 1975, the Congress passed Public Law 94-168 known as the Metric Conversion Act of 1975. This act declares that the SI (International System of Units) will be this country's basic system of measurement and establishes the United States Metric Board which is responsible for the planning, coordination, and implementation of the Nation's voluntary conversion to SI.

The Departmental Manual Release No. 1767, dated June 12, 1975, contained the initial Department of the Interior metric conversion policy. In accordance with this policy, an E&R (Engineering and Research) Center Metric Committee was established February 24, 1976, to represent the USBR (Bureau of Reclamation) in the formulation of metrication procedures, evaluation of training aids, and the general application of the Department's metrication policy. In addition to the E&R Center Metric Committee, each of the seven Bureau regions has appointed a regional coordinator who is responsible for introducing the Bureau's SI metric policies and procedures in the regional and project offices. Overall coordination of the USBR's metrication effort is the responsibility of Reclamation's Metric Conversion Officer in the Office of the Commissioner, Washington, D.C. The Departmental Manual Release No. 1767 has been superseded by Release No. 2056, dated January 19, 1978.

The Metric Manual is based upon practices adopted by the ANMC (American National Metric Council), NBS (National Bureau of Standards), ISO (International Organization for Standardization), ANSI (American National Standards Institute), ASTM (American Society for Testing and Materials), and the ASME (American Society of Mechanical Engineers). Additional sources of information and guidance are listed in the bibliography. The style chapter also considered the USBR Standard Abbreviations List and the GPO Style Manual where applicable.

The material presented herein is a compilation of information and data applicable to the Bureau of Reclamation's metric conversion process. Lawrence D. Pedde is the principal author, coordinator, and editor of this material. Significant contributions were also made by coauthors Warren E. Foote, LeRoy F. Scott, Danny L. King, and Dave L. McGalliard.

Also, special appreciation is extended to Mr. Louis Sokol, President, U.S. Metric Association, who provided significant literature and personal assistance in the compilation of this manual, helped establish the Bureau of Reclamation training program, and reviewed the manuscript for conformance to accepted SI practices.

The metric guidelines and policies described in this manual have been developed for use by the Bureau of Reclamation; no warranty as to the accuracy, usefulness, or completeness is expressed or implied.

The information contained in this publication regarding contractors or commercial products may not be used for advertising or promotional purposes and is not to be construed as an endorsement by the Bureau of Reclamation.

CONTENTS

	Page
Preface	v
Introduction	xvii

Chapter I—The International System of Units—A Description

Introduction	1
Base units	1
Supplementary units	3
Derived units	4
Prefixes	10
Symbols	11

Chapter II—SI Style and Editorial Guidelines

Introduction	13
Capitalization	13
Unit names	13
Symbols	13
Prefixes	14
Punctuation	15
Period	15
Decimal marker	15
Plurals	15
Unit names	15
Symbols	16
Grouping of digits	16
Numerals	16
Four-digit numbers	16
Miscellaneous numbers	18
Spacing—Symbols and names	18
SI multiples and submultiples	18
Unit measure	18
Unit modifier	18
Algebraic symbols	19

CONTENTS—Continued

Page

Superscripts—Squares, cubes, etc. 19

    Symbols 19
    Unit names 20

Compound units 20
Spelling 22
Prefix usage 23
Symbols for angles and temperature 24
Symbol modifiers 25
Algebraic equations and formulas 25
Presentation of chemical elements 26
Abbreviated summary of SI style notes 27

Chapter III—Featured Units

Introduction 29
Units and quantities 29

    Area 29
    Angles 30
    Common fractions 30
    Energy 30
    Force, weight, and mass 31
    Linear measure 34
    The liter 34
    The metric ton 35
    Pressure and stress 36
    Relative density 37
    The steradian 37
    Temperature 41
    Time 42
    Torque 43

USBR preferred units of measure 43
Physical and mental reference guides for SI units 63

    The 10-11-12-13 relationship 63
    Water relationship 63
    References for linear measurements 63
    References for area measurements 64
    References for volume measurements 64
    References for mass measurements 65
    References for temperatures 65

CONTENTS—Continued

Page

Chapter IV—USBR Engineering Applications

Introduction 67
Measurements and dimensioning in technical documents 67

Specifications 67
Publications 68

Contract specifications 68

Tolerances for concrete construction 73
Finishes and finishing 83
Coarse aggregate 89
Sand 95
Polyvinyl-chloride waterstops 98
Rubber waterstops 106
Insulated gypsum wallboard system 114

Materials 118

Electrical conductors 118
Metric screw threads 118
Pipe dimensions 124

Guidelines for SI design drawings 125

Introduction 125
Format 125
Units for design drawings 126
Drafting practices 129
Surveying and mapping 149

International size paper 151
Reading metric gages 154

Micrometers 154
Vernier scale 156
Metric dial calipers 156

Soil classification 156

Sieve sizes 156
Particle classification 163

CONTENTS—Continued

Page

Use of the SI (International System of Units) in
data processing 163

    Introduction 163
    Character set 163
    Rules to represent SI units in print output 163
    Representation of units 165
    Prefixes 166
    Decimal marker 167
    Grouping of digits 167

Chapter V—Unit and Formulae Conversions

Introduction 169
Dimensions and dimensional analysis 169
Guidelines for conversion and rounding of numerical values 174

    Definitions 174
    Conversion and significant digits 175
    Manipulation of numerical data 176
    Rounding of data 178
    Rounding of converted values without tolerances 178
    Rounding of converted values with tolerances 179
    Converting temperatures 182
    The conversion procedure 182

Rationalized metric values 183
Scientific notation 185
Organization and use of the conversion tables 187

    Conversion tables 188

Chapter VI—Engineering Problems and Formulae

Introduction 221
Mechanics and dynamics 221

    Mass and force problems 221
    Active cohesionless soil pressure 223
    Drop hammer loading 224
    Pump power 224
    Vibration 225

CONTENTS—Continued

	Page
Structures	226
Beam deflection	226
Beam shear stress	226
Pipe (beam) stresses	227
Bending moment	228
Concrete beam	229
Prestressed concrete	231
Hydrology	234
Fluid dynamics	234
Flood hydrology	238
Sedimentation	241
Water utilization	242
Irrigation applications	244
Atmospheric water	246
Miscellaneous	247
Bibliography	251
Appendix	257
Glossary	267
Index	271

CONTENTS—Continued

FIGURES

Figure		Page
1-1	Length comparison of the meter and the yard	2
1-2	Mass comparison of the kilogram and the pound	3
1-3	Comparison of the three temperature scales	4
1-4	Illustrated examples of the radian	5
1-5	Illustrated examples of the steradian	6
3-1	American football field—0.535 hectare	29
3-2	Vector comparison of work and torque	31
3-3	Element analysis of spherical sector	39
3-4	Relationship of solid angle to the maximum included plane angle	41
4-1	ISO basic thread profile	120
4-2	Typical thread and tolerance designations	120
4-3	OMFS basic thread profile	124
4-4	USBR SI metric symbol	127
4-5	USBR standard title block	127
4-6	Comparison of multiview projection methods and symbols	128
4-7	Illustrated use of dual dimensioning table	131
4-8	Examples of tolerance formats	135
4-9	Form and position tolerance symbols	136
4-10	Example of position tolerance symbol	137
4-11	The preferred bar scale graduations for scale ratios between 2:1 and 1:5 000 000	139
4-12	Use of bar scales on design drawings	141
4-13	Dimensioning inside and outside diameters	145
4-14	The dimensioning and interpretation of holes	147
4-15	The dimensioning and interpretation of a counter-bored, tapped hole	148
4-16	Examples of various tapped hole designations and the interpretation	149
4-17	Plotting the transition curve	150
4-18	Typical metric micrometer—graduated in hundredths of a millimeter	154
4-19	A vernier scale for a micrometer graduated in thousandths of a millimeter	155
4-20	Metric calipers with vernier scales	157
4-21	Dial face metric caliper	159
4-22	Test pit log sheet	162
5-1	Excerpted portion of length conversion table	187
6-1	Concrete beam in axial compression with a bending moment	229
6-2	Concrete I-beam with reinforcing tendon	232
6-3	Nomograph for Manning's equation	239

CONTENTS—Continued

TABLES

Table		Page
1-1	SI base and supplementary units	2
1-2	SI derived units expressed in base and supplementary units	7
1-3	SI derived units with approved special names	8
1-4	SI derived units expressed by means of derived units having special names	9
1-5	SI unit prefixes	10
2-1	Examples of lowercased and capitalized symbols	14
2-2	Modified symbols	25
3-1	Elements of mass, force, and force of gravity	32
3-2	Common pressures expressed in SI	36
3-3	Solid angle (ψ in steradians)—plane angle (a in degrees) corresponding values	41
3-4	Common torque values	43
3-5	USBR preferred units—Space and time	45
3-6	USBR preferred units—Periodic and related phenomena	48
3-7	USBR preferred units—Mechanics	49
3-8	USBR preferred units—Heat	52
3-9	USBR preferred units—Acoustics	54
3-10	USBR preferred units—Electricity and magnetism	55
3-11	USBR preferred units—Light and electromagnetic radiations	58
3-12	USBR preferred units—Physical chemistry and molecular physics	59
3-13	USBR preferred units—Nuclear physics and ionizing radiations	60
3-14	USBR preferred units—Water resources engineering	61
4-1	Metric units for bidding schedule items in specifications	69
4-2	Substituted SI metric values in standard specifications paragraphs	72
4-3	Actual customary values for underlined SI metric values for polyvinyl-chloride waterstops	105
4-4	Actual customary values for underlined SI metric values for rubber waterstops	113
4-5	Actual customary values for underlined SI metric values for gypsum wallboard	118

CONTENTS—Continued

TABLES—Continued

Table		Page
4-6	Equivalent metric dimensions of standard electrical conductor sizes	119
4-7	Tolerance grades	121
4-8	Tolerance position symbols	121
4-9	Metric threads for commercial screws, bolts, and nuts	122
4-10	Preferred tolerance classes	123
4-11	Nominal SI metric pipe sizes	125
4-12	AWWA recommended units for pipe dimensions	126
4-13	Dimensional units for design drawings	129
4-14	Hot-rolled structural steel shape designations	134
4-15	ISO standard metric drafting scale ratios	137
4-16	Customary versus ISO drafting scale ratios	138
4-17	ISO drafting symbols	146
4-18	Map and site plan scale ratios	151
4-19	A-series paper sizes	152
4-20	B-series paper sizes	152
4-21	C-series paper sizes	153
4-22	Mill paper widths	153
4-23	U.S. and ISO standard test sieves and soil gradations	160
4-24	Visual classification—Description of particle size	163
4-25	SI base units	165
4-26	SI supplementary units	165
4-27	Derived SI units with special names	166
4-28	Other units	166
4-29	Representation of prefixes	167
5-1	Basic dimensional units	169
5-2	Compound dimensional units	170
5-3	Definitions of terminology	174
5-4	General rounding of data	180
5-5	Rounding converted tolerances in inches to millimeters	181
5-6	Step procedure for determining decimal positions when converting precision unit dimensions and tolerances	181
5-7	Conversion of temperature tolerances	182
5-8	Examples of decimal series	184
5-9	Renard series	185
5-10	Incremental measures of length	186

CONTENTS-Continued

TABLES-Continued

Tables		Page
5-11	Acceleration	189
5-12	Area	190
5-13	Chemical concentration	191
5-14	Density—Mass capacity	192
5-15	Electricity—Magnetism	193
5-16	Electromagnetic radiation	195
5-17	Energy: Work—Thermal—Electrical	196
5-18	Energy content of fuels	197
5-19	Energy per area time	198
5-20	Flow	199
5-21	Force	201
5-22	Force per length	201
5-23	Frequency	201
5-24	Grain conversions	202
5-25	Heat	203
5-26	Hydraulic conductivity—Permeability	207
5-27	Illumination	208
5-28	Inertia	209
5-29	Length	210
5-30	Linear density	211
5-31	Load concentration	211
5-32	Mass	212
5-33	Plane angles	213
5-34	Power	214
5-35	Pressure—Stress	214
5-36	Temperature	216
5-37	Time	217
5-38	Torque—Bending moment	217
5-39	Transmissivity	218
5-40	Velocity—Speed	218
5-41	Viscosity	219
5-42	Volume—Capacity	220
A-1	Listing of accepted non-SI units	259
A-2	List of units not to be used with SI	259
A-3	SI units named after scientists	260
A-4	Mathematical and physical constants	260
A-5	List of abbreviations and symbols for SI design drawings	261
A-6	Chemical elements and their symbols	266

INTRODUCTION

A new modern metric system is being adopted worldwide. This new system of measure is called the International System of Units, officially abbreviated SI in all languages. The SI was established by the 11th CGPM (General Conference of Weights and Measures) in 1960; it is intended as a basis for worldwide standardization of measurement units.

The SI is standardized for international use through the ISO (International Organization for Standardization). To achieve common usage and reduce misunderstanding, the standards established by ISO must be followed. Changes to ISO standards can be petitioned through the American National Standards Institute, the U.S. representative/member of the ISO committee.

The Bureau of Reclamation is now progressing towards almost exclusive use of SI. During the early transition period, it was found that several non-SI units of measure will be required; this is the same experience encountered in other countries and by U.S. technical societies. To protect the SI from degradation and to maintain uniformity in presenting technical achievements to an international readership, this publication establishes style and usage guidelines to be followed by Bureau of Reclamation personnel in their transition to the new metric system.

In addition to the style guidelines, the basics of SI are presented, along with the preferred SI and allowable non-SI units, metric conversion techniques, and examples of engineering problems. All of these are required to achieve uniform usage, facilitate comprehension, and reduce errors.

The Bureau of Reclamation is scheduled to complete the conversion to SI by June 1980. With several exceptions, the conversion process presently underway is the "soft" conversion phase. The "soft" conversion involves little more than showing metric equivalents in all new and revised Bureau correspondence, publications, and reports. There are several new construction projects currently being initiated which will use SI units of measure wherever possible and practicable. In these projects, the first step in the transition to SI will be metric surveys followed by metric drawings and specifications; these are the initial steps of a "hard" conversion process.

The speed at which a complete "hard" conversion will be accomplished is essentially in the hands of American industry. It is the policy of the Bureau not to force the use of SI, but to follow the "Rule of Reason." This rule simply states that metric hardware will be purchased and used only when it is commercially available and economically feasible. The Bureau will not lead industry but will keep pace with it.

Chapter I

THE INTERNATIONAL SYSTEM OF UNITS—A DESCRIPTION

Introduction

The 11th CGPM (General Conference on Weights and Measures) established the SI (International System of Units) to serve as a worldwide standard of measurement units. The system offers numerous advantages in world trade, international relations, teaching, and scientific work.

The SI units are divided into three classes: (1) base units, (2) derived units, and (3) supplementary units. The seven base units were selected by the CGPM; they are well defined and considered to be dimensionally independent.

The second class of SI units consists of the derived units; these are units formed by algebraic relations between the base units, the supplementary units, other derived units, specially named derived units, or a combination of these.

The two supplementary units were added by the 11th CGPM. The CGPM declined to designate these as base or derived units, thus the existence of the third category. The classification of SI units into these classes is somewhat arbitrary; a unit's classification is not really important to the use and understanding of the system.

The units in these three classes form a coherent measurement system. A system of units is coherent if the product or quotient of any multiple-unit quantity in the system is the unit of the resultant quantity. Examples of this are 1 newton meter (N·m) equals 1 joule (J) and 1 newton per 1 square meter (N/m^2) equals 1 pascal (Pa).

The metric system is based upon the decimal system, much like our monetary system; all that is required is to think in terms of 10. The metric system does not have 12 to 1, 36 to 1, 5280 to 1, etc., relationships between its units, only powers of 10; this makes the metric system simple and easy to use.

Base Units

The first class of SI units is the base units. Table 1-1 lists the seven basic physical quantities and the name and symbol of the corresponding SI unit of measure. A definition and discussion of each SI unit follow.

Meter.—The meter is the length equal to 1 650 763.73 wave lengths in vacuum of the orange-red line of the spectrum of the krypton-86 atom (radiation corresponding to the transition between the $2p_{10}$ and $5d_5$ levels).

A common comparative reference for the meter is the yard; the meter is approximately 10 percent longer than the yard (fig. 1-1). Examples of certain lengths or distances measured in meters include: (1) a football field, goal line to goal line, is slightly more than 91 meters; (2) the height of the average American male is a little less than 1.8 meters; (3) the distance between the bases in baseball is about 27 meters; and (4) the length of a full-size bed is 2 meters.

Table 1-1.—*SI base and supplementary units*

Quantity	Name	Symbol
BASE UNITS		
Length	meter	m
Mass	*kilogram	kg
Time	second	s
Electric current	ampere	A
Thermodynamic temperature	†kelvin	K
Amount of substance	mole	mol
Luminous intensity	candela	cd
SUPPLEMENTARY UNITS		
Plane angle	radian	rad
Solid angle	steradian	sr

* The kilogram is the only base unit that has a prefix.

† Although the SI unit of thermodynamic temperature is the kelvin, the Celsius scale is the commonly used scale for temperature measurements. The thermodynamic scale, or the Kelvin scale, is used more in scientific work. A Celsius temperature (t_c) is related to a temperature on the Kelvin scale (t_k) by:

$$t_c = t_k - 273.15$$

Kilogram.—The standard for the unit of mass, the kilogram, is a cylinder of platinum-iridium alloy kept by the International Bureau of Weights and Measures in Sevres, France. A duplicate cylinder is in the custody of the National Bureau of Standards and serves as the mass standard for the United States. This is the only base unit still defined by an artifact.

Comparing the kilogram to our present measure for mass, the kilogram is approximately 2.2 times larger than the pound (mass), see figure 1-2.[1] Examples of masses measured in kilograms include: (1) bags of flour in 2- and 4-kilogram bags instead of 5- and 10-pound bags, and (2) the mass of a professional football defensive lineman is about 115 kilograms.

```
┌─────────────────────┐
│   1  YARD           │        = 0.9144  meter
└─────────────────────┘
```

```
┌─────────────────────┐
│   1  METER          │
└─────────────────────┘
```

Figure 1-1.—Length comparison of the meter and the yard. 40-D-6330

---

[1] Mass, force, and weight are discussed in greater detail in chapter III.

CHAPTER 1—THE INTERNATIONAL SYSTEM OF UNITS

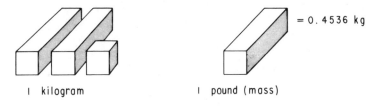

Figure 1-2.—Mass comparison of the kilogram and the pound. 40-D-6331

Second.—The second is the duration of 9 192 631 770 periods of the radiation corresponding to the transition between the two hyperfine levels of the ground state of the cesium-133 atom.

The second was previously defined as 1/86 400 of the mean solar day. The solar day was found to vary due to the changes in the Earth's rotation speed, so a new and more exact standard was selected.

Ampere.—The ampere is that constant current which, if maintained in two straight parallel conductors of infinite length, of negligible circular cross section, and placed 1 meter apart in vacuum, would produce between these conductors a force equal to 2×10^{-7} newton per meter of length.

Kelvin.—The kelvin unit of thermodynamic temperature is the fraction 1/273.16 of the thermodynamic temperature of the triple point of water.

Most temperatures will be measured in degrees Celsius. The kelvin temperatures will be limited to scientific use. See figure 1-3 for the general relationships among the Fahrenheit, Celsius, and Kelvin scales.

Mole.—The mole is the amount of substance of a system which contains as many elementary entities (atoms, molecules, ions, electrons, other particles, or specified groups of such particles) as there are atoms in 0.012 kilogram of carbon-12.

Candela.—The candela is the luminous intensity, in the perpendicular direction, of a surface of 1/600 000 square meter of a black body at the temperature of freezing platinum under a pressure of 101 325 pascals (newtons per square meter).

Supplementary Units

Another class of SI units is the supplementary units; this class consists of the radian and steradian.

Table 1-1 also lists the two supplementary units, with symbols, and figures 1-4 and 1-5 provide graphic examples of the radian and steradian. A definition and discussion of these two units follow.

Radian.—The radian is the plane angle with its vertex at the center of a circle that is subtended by an arc equal in length to the radius.

One radian equals 57.2958°; the angle of a complete circle equals 2π radians ($\simeq$ 6.2832 rad).

The use of the radian will be limited. The Bureau of Reclamation will continue to express plane angles in degrees. Angles expressed in radians will be used primarily in scientific and engineering equations.

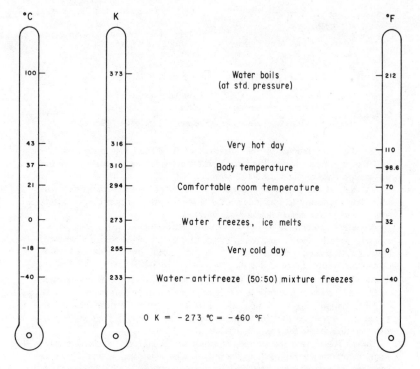

Figure 1-3.—Comparison of the three temperature scales. 40-D-6332

Steradian.—The steradian is the solid angle with the vertex at the center of a sphere that is subtended by an area of the spherical surface equal to that of a square having sides equal in length to the radius.

A sphere equals 4π steradians ($\simeq$ **12.5664** sr).

Plane and solid angles are discussed in greater detail in chapter III.

Derived Units

The largest class of SI units, the derived units, is formed by combining base, supplementary, and other derived units according to the algebraic relations linking the corresponding quantities. When two or more units expressed in base or supplementary units are multiplied or divided as required to obtain derived quantities, the result is a unit value. A numerical constant is not introduced, thus maintaining a coherent system.

Several derived units have been given special names and symbols which may themselves be used to express other derived units in a simpler way than in terms of the base or supplementary units. For example, joule is the name given to the algebraically equal units of newton meter (N·m) and kilogram meter

CHAPTER 1—THE INTERNATIONAL SYSTEM OF UNITS

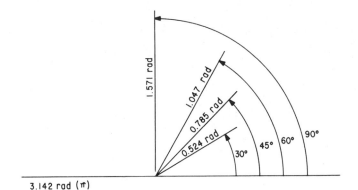

Examples of Plane Angles

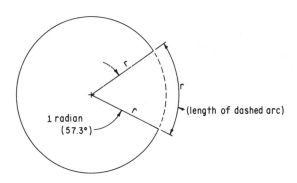

Illustrated Definition

Figure 1-4.—Illustrated examples of the radian. 40-D-6333

squared per second squared ($kg \cdot m^2/s^2$). Examples of some derived units are listed in tables 1-2, 1-3, and 1-4.

Listed below are: (1) the quantities which have been given special names, (2) the special name, and (3) the definition of the name. Table 1-3 supplements this list with symbol representations.

Electrical capacitance.—The farad is the capacitance of two plates between which there exists a potential difference of 1 volt when each plate possesses an equal and opposite charge of electricity equal to 1 coulomb.

Electrical conductance.—The siemens is the electrical conductance of a conductor in which a current of 1 ampere is produced by an electrical

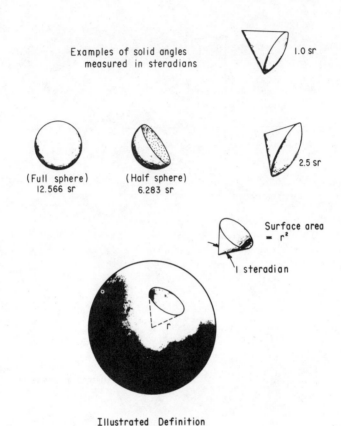

Figure 1-5.—Illustrated examples of the steradian. 40-D-6334

potential difference of 1 volt. The siemens replaces the mho on a one-to-one basis.

Electrical inductance.—The henry is the inductance of a closed circuit in which an electromotive force of 1 volt is produced when an electric current in the circuit varies uniformly at a rate of 1 ampere per second.

Electric potential difference.—The volt (unit of electric potential difference and electromotive force) is the difference of electric potential between two points of a conductor carrying a constant current of 1 ampere, when the power dissipated between these points is equal to 1 watt.

Electrical resistance.—The ohm is the electric resistance between two points of a conductor when a constant difference of potential of 1 volt, applied between these two points, produces in this conductor a current of 1 ampere, this conductor not being the source of any emf (electromotive force).

Table 1-2.—*SI derived units expressed in base and supplementary units*

Quantity	Unit	Symbol
Acceleration	meter per second squared	m/s^2
Angular acceleration	radian per second squared	rad/s^2
Angular velocity	radian per second	rad/s
Area	square meter	m^2
Concentration (amount of substance)	mole per cubic meter	mol/m^3
Current density	ampere per square meter	A/m^2
Density, mass density	kilogram per cubic meter	kg/m^3
Luminance	candela per square meter	cd/m^2
Magnetic field strength	ampere per meter	A/m
Radiance	watt per square meter steradian	$W/(m^2 \cdot sr)$
Radiant intensity	watt per steradian	W/sr
Rotational frequency	revolution per second	r/s
Specific volume	cubic meter per kilogram	m^3/kg
Speed, velocity	meter per second	m/s
Thermal flux density	watt per square meter	W/m^2
Volume	cubic meter	m^3
Wave number	1 per meter	*m^{-1}

*To use $1/m$, instead of m^{-1}, may cause confusion; it may be read as liter per meter.

Energy and work.—The joule is the work done when the point of application of a force of 1 newton is displaced a distance of 1 meter in the direction of the force.

Force.—The newton is that force which, when applied to a body having a mass of 1 kilogram, gives it an acceleration of 1 meter per second squared.

Frequency.—The hertz is a frequency of 1 cycle per second.

Illuminance.—The lux is the illuminance produced by a luminous flux of 1 lumen uniformly distributed over a surface of 1 square meter.

Luminous flux.—The lumen is the luminous flux emitted in a solid angle of 1 steradian by a point having a uniform intensity of 1 candela.

Magnetic flux.—The weber is the magnetic flux which, linking a circuit of one turn, produces in it an electromotive force of 1 volt as it is reduced to zero at a uniform rate of 1 second.

Magnetic flux density.—The tesla is the magnetic flux density given by a magnetic flux of 1 weber per square meter.

Power.—The watt is the power which gives rise to the production of energy at the rate of 1 joule per second.

Pressure or stress.—The pascal is the pressure or stress of 1 newton per square meter.

Quantity of electricity.—The coulomb is the quantity of electricity transported in 1 second by a current of 1 ampere.

Absorbed dose.—The gray is the energy absorbed by a mass from an ionizing radiation; the gray is defined as 1 joule per kilogram. The gray replaces the rad.

Table 1-3.—SI derived units with approved special names

Quantity	Special name	Symbol	Other units	Expressed in terms of: SI base and supplementary units (equivalent expressions)	
Absorbed dose	gray	Gy	J/kg	$m^2 \cdot s^{-2}$	m^2/s^2
Activity (radioactive)	becquerel	Bq		s^{-1}	
Electric capacitance	farad	F	C/V	$m^{-2} \cdot kg^{-1} \cdot s^4 \cdot A^2$	$s^4 \cdot A^2/(kg \cdot m^2)$
Electric charge, quantity of electricity	coulomb	C		$A \cdot s$	
Electric conductance	siemens	S	A/V	$m^{-2} \cdot kg^{-1} \cdot s^3 \cdot A^2$	$s^3 \cdot A^2/(kg \cdot m^2)$
Electric inductance	henry	H	Wb/A	$m^2 \cdot kg \cdot s^{-2} \cdot A^{-2}$	$m^2 \cdot kg/(s^2 \cdot A^2)$
Electric potential, potential difference, electromotive force	volt	V	W/A	$m^2 \cdot kg \cdot s^{-3} \cdot A^{-1}$	$m^2 \cdot kg/(s^3 \cdot A)$
Electric resistance, impedance	ohm	Ω	V/A	$m^2 \cdot kg \cdot s^{-3} \cdot A^{-2}$	$m^2 \cdot kg/(s^3 \cdot A^2)$
Energy, work, quantity of heat	joule	J	N·m	$m^2 \cdot kg \cdot s^{-2}$	$m^2 \cdot kg/s^2$
Force	newton	N		$m \cdot kg \cdot s^{-2}$	$m \cdot kg/s^2$
Frequency	hertz	Hz		s^{-1}	
Illuminance	lux	lx	lm/m^2	$m^{-2} \cdot cd \cdot sr$	$cd \cdot sr/m^2$
Luminous flux	lumen	lm		$cd \cdot sr$	
Magnetic flux	weber	Wb	V·s	$m^2 \cdot kg \cdot s^{-2} \cdot A^{-1}$	$m^2 \cdot kg/(s^2 \cdot A)$
Magnetic flux density	tesla	T	Wb/m^2	$kg \cdot s^{-2} \cdot A^{-1}$	$kg/(s^2 \cdot A)$
Power radiant flux	watt	W	J/s	$m^2 \cdot kg \cdot s^{-3}$	$m^2 \cdot kg/s^3$
Pressure, stress, modulus	pascal	Pa	N/m^2	$m^{-1} \cdot kg \cdot s^{-2}$	$kg/(m \cdot s^2)$

Table 1-4.—SI derived units expressed by means of derived units having special names

Quantity	Name	Symbol		Expressed in base units (equivalent expressions)
Dynamic viscosity	pascal second	Pa·s	$m^{-1} \cdot kg \cdot s^{-1}$	$kg/(m \cdot s)$
Electric charge density	coulomb per cubic meter	C/m³	$m^{-3} \cdot s \cdot A$	$s \cdot A/m^3$
Electric field strength	volt per meter	V/m	$m \cdot kg \cdot s^{-3} \cdot A^{-1}$	$m \cdot kg/(s^3 \cdot A)$
Electric flux density	coulomb per square meter	C/m²	$m^{-2} \cdot s \cdot A$	$s \cdot A/m^2$
Energy density	joule per cubic meter	J/m³	$m^{-1} \cdot kg \cdot s^{-2}$	$kg/(m \cdot s^2)$
Heat capacity, entropy	joule per kelvin	J/K	$m^2 \cdot kg \cdot s^{-2} \cdot K^{-1}$	$m^2 \cdot kg/(s^2 \cdot K)$
Heat flux density, irradiance	watt per square meter	W/m²	$kg \cdot s^{-3}$	kg/s^3
Molar energy	joule per mole	J/mol	$m^2 \cdot kg \cdot s^{-2} \cdot mol^{-1}$	$m^2 \cdot kg/(s^2 \cdot mol)$
Molar entropy, molar heat capacity	joule per mole kelvin	J/(mol·K)	$m^2 \cdot kg \cdot s^{-2} \cdot K^{-1} \cdot mol^{-1}$	$m^2 \cdot kg/(s^2 \cdot K \cdot mol)$
Moment of force	newton meter	N·m	$m^2 \cdot kg \cdot s^{-2}$	$m^2 \cdot kg/s^2$
Permeability (electric)	henry per meter	H/m	$m \cdot kg \cdot s^{-2} \cdot A^{-2}$	$m \cdot kg/(s^2 \cdot A^2)$
Permittivity	farad per meter	F/m	$m^{-3} \cdot kg^{-1} \cdot s^4 \cdot A^2$	$s^4 \cdot A^2/(m^3 \cdot kg)$
Specific energy	joule per kilogram	J/kg	$m^2 \cdot s^{-2}$	m^2/s^2
Specific heat capacity, specific entropy	joule per kilogram kelvin	J/(kg·K)	$m^2 \cdot s^{-2} \cdot K^{-1}$	$m^2/(s^2 \cdot K)$
Surface tension	newton per meter	N/m	$kg \cdot s^{-2}$	kg/s^2
Thermal conductivity	watt per meter kelvin	W/(m·K)	$m \cdot kg \cdot s^{-3} \cdot K^{-1}$	$m \cdot kg/(s^3 \cdot K)$

Activity.—The becquerel is the activity of radionuclides having one nuclear transition per second. The becquerel replaces the curie.

Prefixes

The prefixes for SI units, listed in table 1-5, are used to form decimal multiples and submultiples of the SI units. Only one multiplying prefix is applied at one time to a given unit. The symbol of a prefix is considered to be combined with the unit symbol to which it is directly attached.

The combinations formed by attaching the prefixes with base, supplementary, or derived units technically must be considered "multiples and submultiples" of SI units. However, in common usage, all such units are considered to be SI units.

Although the guidelines and requirements of the situation will determine the actual prefixes used, it is believed that for the majority of USBR work, four prefixes will satisfy the needs. These prefixes are mega (10^6), kilo (10^3), milli (10^{-3}), and micro (10^{-6}).

The selection of prefixes is **not** restricted to these. These four prefixes are multiples of 1000, which follows the basic selection preference. The use and selection of prefixes are discussed more in chapter II.

Table 1-5.—*SI unit prefixes*

Multiplication factor	Prefix	Symbol	Pronunciation	
1 000 000 000 000 000 000 = 10^{18}	exa	E	ex′ a	x-ah
1 000 000 000 000 000 = 10^{15}	peta	P	pet′ a	pet-ah
1 000 000 000 000 = 10^{12}	tera	T	ter′ a	terr-ah
1 000 000 000 = 10^9	giga	G	jig′ a	jig-ah
1 000 000 = 10^6	mega	M	meg′ a	megg-ah
1 000 = 10^3	kilo	k	kil′ o	kill-oh
100 = 10^2	*hecto	h	hec′ to	heck-toe
10 = 10^1	*deka	da	dek′ a	deck-ah
0.1 = 10^{-1}	*deci	d	dec′ i	dess-ee
0.01 = 10^{-2}	*centi	c	cent′ i	sent-ee
0.001 = 10^{-3}	milli	m	mill′ i	mill-ee
0.000 001 = 10^{-6}	micro	μ	mic′ ro	mike-ro
0.000 000 001 = 10^{-9}	nano	n	nan′ o	nan-oh
0.000 000 000 001 = 10^{-12}	pico	p	pic′ o	peek-oh
0.000 000 000 000 001 = 10^{-15}	femto	f	fem′ to	fem-toe
0.000 000 000 000 000 001 = 10^{-18}	atto	a	att′ o	at-oh

*Except for the nontechnical use of centimeter and in special area and volume measures, the use of these prefixes should be avoided when possible.

Symbols

The short forms of SI unit names and prefixes are called symbols; it is incorrect to refer to them as abbreviations or acronyms. In the previous tables, the symbol representations have been presented for all the associated unit names and prefixes.

Almost all SI symbol are written in upright roman letters regardless of the lettering style used for the associated material; the two exceptions are the Greek mu (μ, for micro) and omega (Ω, for ohm). Maintain the clear distinction between SI symbols and quantity symbols (e.g., M for mass). Quantity symbols are written in italic or sloping letters.

More information on the use of symbols is presented in chapters II and IV.

Chapter II
SI STYLE AND EDITORIAL GUIDELINES

Introduction

This chapter presents the basic rules to be used in writing SI units. Some basic items such as capitalization, punctuation, and spelling are discussed as are some new concepts concerning number groupings and prefix selection. The guidelines are intended to provide the groundwork required to use a new language in a clear and concise manner. Deviations from these guidelines will most likely result in incorrect information or misinterpretation.

It is emphasized that these guidelines generally should not be applied to the U.S. customary system. Documents exclusively using the U.S. customary system units should continue to be treated as they have in the past. However, documents which use a dual units method of presentation are to follow the SI style rules for both systems. This includes the use of spaces instead of commas in numbers, the use of product dots in compound symbols instead of hyphens, and other appropriate format rules.

Capitalization

Unit Names

The SI unit names, including prefixes, are not capitalized when used within a sentence, except the first letter is capitalized when used as the first word of a sentence. The unit names are treated as common nouns; when part of a title, write the SI unit names in the same format as the rest of the title.

Note that in the term "degree Celsius," "degree" is considered to be the unit name, modified by the adjective "Celsius." Degree is written in lowercase letters but Celsius is always capitalized. Note that the kelvin unit is lowercase, but when describing the scale, it is the Kelvin scale, giving recognition to the originator.

Symbols

The short forms for metric units are called symbols. Unit symbols are written with lowercase letters, except that the first letter is capitalized when the name of the unit is derived from the name of a person. Table 2-1 lists some SI units and their symbols.

A new exception to this rule is the capital letter L symbol for liter. The official SI symbol for liter remains lowercase "el" (l); but, because there is no clear difference between the lowercase "el" and the number one on most typing equipment, the capital L presents a distinct and logical alternative. The NBS, ANMC, ASTM, and several other leading U.S. technical organizations have instituted the use of L; the Bureau of Reclamation will follow this policy because it reflects the dominant trend within the United States.

Table 2-1.—*Examples of lowercased and capitalized symbols*

Lowercase symbols		Capitalized symbols		
meter	m	ampere	A	A. M. Ampere
gram	g	kelvin	K	Wm. Thompson, 1st Baron Kelvin
second	s	pascal	Pa	Blaise Pascal
candela	cd	newton	N	Isaac Newton
mole	mol	degree Celsius	°C	Anders Celsius
radian	rad	hertz	Hz	Heinrich R. Hertz

It is preferred that symbols be used only when writing a measurement value or when the unit name is exceptionally complex; otherwise, use the unit name. Symbols are not to be used to start a sentence; therefore, no confusion should occur regarding a possible change in capitalization. The following are examples of these guides:

> Correct—The length was measured in meters; the final measure was 6.1 m. Hydraulic conductivity can be given in $m^3/(m^2 \cdot s)$ or the simplified form of meters per second.
> Incorrect—G is the common measure for small masses. 6 g was the smallest mass found. Length dimensions are given in km.

Do not use unit symbols in capitalized material such as a title. Whenever circumstances would require incorrect capitalization of unit symbols, it is recommended that unit names be used instead.

Prefixes

All prefix names, their symbols, and pronunciation are listed in table 1-5. Notice that the first five prefixes at the top of the list have their symbols shown as capital letters. All other prefix symbols are lowercase letters.

All prefix names (plus SI unit names) are written in lowercase letters when fully written out in a sentence. The first letter of the prefix is written in capital letters at the beginning of a sentence; when part of a title, write the prefix and unit name in the same format as other common nouns.

The importance of following the precise use of capital and lowercase letters is shown by the following examples:

> G for giga; g for gram
> K for kelvin; k for kilo
> M for mega; m for milli
> N for newton; n for nano
> T for tera; t for metric ton

Punctuation

Period

When symbols are used for metric units, a period is **not** used after the symbol, except at the end of a sentence.

Example: The 100-m-long rope was found to have a mass of 200 kg.

On typing equipment and computers which have a limited character set, namely, no product dot, it is permissible to use a period in the same space that a product dot would be placed. This practice is considered acceptable on an interim basis until the product dot becomes a standard character. The raised product dot is preferred and should be used where practicable.

Examples:

	Preferred	Acceptable
newton meter	N·m	N.m
pascal second	Pa·s	Pa.s

Decimal Marker

A dot on the line (a period) is used as a decimal marker. The decimal marker is to be bold and is to occupy one full letter space. In numbers having a value less than one, a zero is written before the decimal marker to prevent the possibility that a faint decimal marker will be overlooked.

Example: The oral expression "one-half kilogram of cement" is written "0.5 kg of cement."

Plurals

Unit Names

When written in full, the names of metric units are made plural (i.e., adding an "s") when appropriate. Values between +1 and −1 are always singular; for other values, add the "s."

Certain SI units are both singular and plural; examples of this are lux, hertz, and siemens. These unit names do not change with the magnitude of the measure.

Examples: 1 meter   15 meters
           0.1 kilogram   29 kilograms

The frequency varied ± 0.5 hertz about a midpoint of 60 hertz.

A measurement value of zero provides a point of discontinuity in what persons write and what they say. The rule for using the singular form of a unit name applies to a zero value, but that is not the way the units are usually read. Examples of this are 0 °C or 0 N; these would be recognized as singular but would be read as plurals, namely, zero degrees Celsius and zero newtons, respectively.

Using symbols for measurement values is preferred anyway, but for zero measures, the use of the symbol prevents the possible incorrect use of a plural unit name when a singular form is required.

Symbols

Symbols for SI units are never plural. Never add an "s" to indicate a plural.

Examples: 1.7 m   1 m   0.6 m
          -30 °C   0 °C   100 °C

Grouping of Digits[1]

Numerals

All numbers are composed of individual digits, 0 through 9. With the exception of the special consideration given the "four-digit number" (discussed in the following section), all numbers are separated into groups of three digits on each side of the decimal marker. Do **not** use a comma to separate the groups of three digits. A space is left between the number groups instead of the traditional comma; the purpose of this is to avoid the confusion with other countries where the comma is used for the decimal marker.

A space is not left between the digits and the decimal marker.

Four-digit Numbers

Four-digit numbers are given special consideration and are treated differently depending upon the context in which they are used, text or tabular format. In text material, all numbers having four or less digits on either side of the decimal marker are to be written with no spaces. Thus, an eight-digit number, four digits on each side of the decimal marker, would be written 1234.5678.

Normally in tabular listings, all numbers use the three-digit grouping and spaces format. Adopting this format for all tables is preferred as it eliminates possible format errors. However, it is not uncommon to have a column of numbers, none of which have more than four digits on either side of the decimal marker; in cases such as this, it is permissible (but not required) to write the numbers without spaces even though the numbers are in tabular format.

---

[1] Drafting guidelines follow these rules, but in ADP, all numbers are written in block numbers. See the appropriate sections in chapter IV.

CHAPTER 2—SI STYLE AND EDITORIAL GUIDELINES

Please note there are two general types of tables, one in which the column of numbers are written with the decimal position in vertical alinement and one in which the data listed are not uniformly arranged. In the first type, the number groupings and spaces **must be** consistent within the column; variations between columns is permissible. In the latter, the entire table must be consistent, either with spaces used or spaces not used as in all four-digit format.

A number must be treated equally on both sides of the decimal marker; if a number requires spaces on one side of the decimal marker, then the appropriate spacing must also be used on the opposite side of the decimal marker. In text or tabular format, it would be incorrect to write numbers such as 1234.567 89 and 99 131.7778. The correct presentation of these values is 1 234.567 89 and 99 131.777 8, respectively.

Text examples:

Replace	With
4,720,525	4 720 525
246.52875	246.528 75
6,875	6875
0.6875	0.6875
0.67812	0.678 12

Tabular examples:

(Preferred)

Table A

	Mass (kg)	Volume (m^3)
W	1 267	281.1
X	6 125	1 671.8
Y	7 121	7 123.1
Z	8 136	8 112.6

Table B

	Area (ha)	Volume (m^3)	Perimeter (km)
WW	15 010	7 610 100	6 010
XX	2 100	2 710 000	1 215
YY	18 121	9 410 400	7 712
ZZ	6 100	800 500	885

Table C

Mass	5 000 to 6 500 kg
Volume	9 670 m^3
Transport distance	6 800 to 9 500 m

(Acceptable)

Table A'

	Mass (kg)	Volume (m^3)
W	1267	281.1
X	6125	1671.8
Y	7121	7123.1
Z	8136	8112.6

Table B'

	Area (ha)	Volume (m^3)	Perimeter (km)
WW	15 010	7 610 100	6010
XX	2 100	2 710 000	1215
YY	18 121	9 410 400	7712
ZZ	6 100	800 500	885

Table C'

Mass	5000 to 6500 kg
Volume	9670 m^3
Transport distance	6800 to 9500 m

Table D

Mass	5 000 to 6 500 kg	(no acceptable
Volume	17 650 m³	alternative)
Transport distance	6 800 to 12 000 m	

Miscellaneous Numbers

There are certain numbers to which the aforementioned grouping rules do not apply. Figures involving serial numbers and currency are not to be changed, but will continue to be written in their original style.[2] Commas, hyphens, spaces, and other applicable symbols will still be used.

Examples: $21,621.67   (currency)
          163HHC-671226/671228 (part number)
          505-54-9126   (social security number)

Spacing—Symbols and Names

SI Multiples and Submultiples

A space or hyphen is not used between either the prefix and unit name or the prefix symbol and the unit symbol.

Example: kiloampere, kA   megawatt, MW

Unit Measure

When a symbol or name follows a number to which it refers, a space must be left between the number and the symbol or name. The symbols for degree, minute, and second (angular measure) are written without a space between the number and the degree symbol; these are not SI units.

Examples: 455 kHz, 455 kilohertz, 20 mm, 10 N, 36°, 36 °C

Unit Modifier

The prescribed method for writing a SI measurement requires that a space be left between the numeric value and the unit symbol or name. However, when a quantity is used in an adjectival sense, a hyphen is often used between the numeric value and the unit symbol or name; do not use a hyphen with the degree of angle symbol (°) or the degree Celsius symbol (°C).

The general rule for the use of the unit modifier hyphen is as follows:

If a dimension is written before a noun, the quantity is idiomatically hyphened to a singular word. But if the dimension is written after the noun, the

---

[2] The treatment of traditional customary measures will depend upon the associated material. See the last paragraph of this chapter's introduction.

CHAPTER 2—SI STYLE AND EDITORIAL GUIDELINES

figure is not hyphenated to the word, and the word is plural (unless of course, the figure is one or less).

When using the SI language, the use of the unit modifier hyphen is not always desirable. Misinterpretation may result because of other rules governing the writing of SI quantities.

Example: We purchased 120 500-kg beams.

Does this say that the beams purchased had a mass of 120 500 kg or were 120 beams purchased, each having a mass of 500 kg? In instances such as this, it would be prudent to restructure the sentence to avoid the use of the hyphen. It is also recommended that SI compound unit names and symbols not be used with unit modifier hyphens; problems may occur with compound unit names which use hyphens or the division "per," or with compound symbols which use the product dot (·) or the division slash (/). Again, sentence revision is recommended. Reason and sentence clarity should be the determining factors in the use of unit modifier hyphens; refer to section 6 of the GPO Style Manual [19] regarding the use of the unit modifier hyphen.

Examples:

The theodolite was rotated in 5° increments.
We purchased sixty 200-kg beams; previously, we purchased 120 beams having a unit mass of 500 kg.
I bought two 6-kg turkeys.
It is a 35-mm camera; the film in the camera is 35 mm wide.

Algebraic Symbols

Leave a space on each side of signs for multiplication, division, addition, and subtraction. This does not apply to compound symbols which use the slash (/) or the product dot (·).

Examples:

4 km + 2 km = 6 km   6 N x 8 m = 48 N·m
26 N ÷ 3 m^2 = 8.67 Pa   100 W ÷ (10 m x 2 K) = 5 W/(m·K)
10 kg/m^3 x 0.7 m^3 = 7 kg   10 Pa·s, 15 kW·h, 0.7 mg/L

It is recommended that unit names not be used in arithmetic/algebraic equations; use the symbols.

Superscripts—Squares, Cubes, Etc.

Symbols

When writing symbols for metric units requiring an exponent, such as square meter, cubic centimeter, one per second, write the superscript immediately after the symbol.

Examples: 14 square meters   14 m²
           10 cubic centimeters   10 cm³
           1 per second   s⁻¹

Unit Names

When writing compound units, certain factors may involve squares, cubes, etc. Many times it is not readily apparent to which factor the superscript name applies. Under these circumstances, it is advisable to use the symbol representation exclusively, or at least include it in parentheses.

Example: kilogram meter squared per second squared

The correct symbol for this is kg·m²/s²; it could be incorrectly interpreted to be (kg·m)²/s² or (kg·m²/s)².

Compound Units

Compound units are derived as quotients or products of other SI units. For compound units, there are a few basic rules to be followed. These rules are listed below:

1. Avoid mixtures of words and symbols. Do not use a slash (/) as a substitute for "per."[3]

Examples: Do not use kilometer/hour or km per hour in place of kilometer per hour or km/h.

The water leakage rate was 0.5 L/s.

2. Use only one "per" in any combination of metric unit names.

Examples: Use meter per second squared, not meter per second per second.

The word "per" denotes the mathematical process of division. Do not use per when "for each" is meant; for example, the measure of leakage current given in microamperes per 1 kilovolt of line-to-line voltage should be written in microamperes for each kilovolt of line-to-line voltage. In SI, 1 μA/kV equals 1 nanosiemens (nS).

3. Do not mix non-SI metric units with SI metric units.

Examples: Use kg/m³, not kg/ft³.
           Use kg/m, not lb/m.

---

[3] The slash is also referred to as a solidus, virgule, slant, and shilling mark.

CHAPTER 2—SI STYLE AND EDITORIAL GUIDELINES

4. To eliminate the problem of what units and multiples to use, a quantity that constitutes a ratio of two like quantities should be expressed as a percentage, decimal fraction, or a scale ratio.

   Example: The slope of 10 m per 100 m can be expressed as 10 percent, 0.10 or 1:10.

   A strain of 100 μm/m can be converted to 0.01 percent.

5. Use only accepted symbols in expressing SI units.

   Example: The correct SI symbol representation for kilometer per hour is km/h; do not write k.p.h., kph, or KPH because these are understood only in English-speaking countries.

6. Similar to rule 2, do not use more than one slash (/) in any combination of symbols unless parentheses are used to separate them.

   Examples: Write m/s^2, not m/s/s.
   Write W/(m·K) or (W/m)/K, not W/m/K.

   [4] It is acceptable to write $(m^3/d)/m^2$ in place of $m^3/(m^2 \cdot d)$.

7. For most unit names derived as a product, a space (preferred) or a hyphen is used to indicate the relationship, but never use a "product dot." Some compound units are acceptable written as one word; no space or hyphen is required, unless the word needs to be broken.

   Examples: Write newton meter or newton-meter; do not write newton·meter.

   It is correct to write watt hour, watt-hour, or watthour. This also applies to kilowatt hour.

8. For product-derived symbols, use a "product dot" between each individual unit symbol. Presently, there are a few exceptions to this but they may eventually be eliminated. Do not use the product dot as a multiplication symbol for calculations.

   Examples: N·m (newton meter)
   Pa·s (pascal second)
   kW·h or kWh (kilowatt hour)
   Use 7.6 x 6.1 cos ϕ, not 7.6·6.1 cos ϕ

---

[4] Either symbol may be dimensionally simplified to m/d; the more defined hydraulic conductivity measure is presented for illustrative purposes.

Presently, there are instances when the period is used in place of the product dot in compound unit symbols. An additional discussion of this is presented in the section on punctuation.

9. Exercise care when writing compound units which involve powers; the modifier, squared or cubed, should be placed after the unit name to which it applies. For powers larger than three, use only symbols.

In the case of area or volume, the modifier (square or cubic) should be placed before the unit name to which it applies. The use of symbols is always preferred when complicated expressions are involved.

> Examples: Write meter per second squared, not meter per square second. There is no such thing as a square second.
>
> The expression for mass moment of inertia is kilogram meter squared. If this were written kilogram square meter, the symbol would be the same but the physical interpretation would be different.

10. For complicated symbol representations, use parentheses to simplify and clarify.

Spelling

The spelling of five SI units may appear questionable. Three of these units involve the addition of a prefix. These units are megohm (mega + ohm), kilohm (kilo + ohm), and hectare (hecto + are). As can be seen by the combinations shown in parentheses, the final vowel from the prefix is dropped from the spelling and is omitted in the pronunciation. In all other cases where the prefix ends with a vowel and the unit name begins with a vowel, both vowels remain and both are pronounced.

The two other units are meter and liter. There has been considerable controversy as to whether the Government should adopt the "re" spellings (metre and litre) as prescribed by ISO 1000 or continue to use the traditional "er" spellings. The Metric Conversion Act of 1975 (PL 94-168) assigns to the Department of Commerce the responsibility to interpret and modify for the United States the International System of Units. The Assistant Secretary for Science and Technology, U.S. Department of Commerce, has concluded that the "er" spelling of meter and liter is preferable for the United States usage. Any exceptions to this recommendation will be limited to those situations where it is appropriate in international relationships. The Assistant Secretary's recommendation will standardize the spelling of "meter" and "liter" within the U.S. Government; however, many of the large U.S. industrial companies have already opted for the international spelling of "metre" and "litre."

Prefix Usage

Prefixes representing 10 raised to a power in multiples of 3 (i.e., 10^{3n}, n = 0, ±1, ±2, etc.), are preferred. While hecto, deka, deci, and centi are common prefixes, their use should generally be avoided; exceptions include the centimeter for consumer products and hectare for land and water areas.

The prefix selection scheme of choosing only multiples of 1000 does not apply to areas and volumes when the meter is the base unit involved.[5] All prefixes are available for use; this helps avoid unit intervals which have relative magnitudes of 10^6 for area and 10^9 for volume.

In specific areas of use, in a table of values for the same quantity, or in a discussion of related values within a given context, a common unit multiple should be used even when some of the numerical values may require up to five or six digits before the decimal marker. The impact of this may be minimized by using scientific notation.

Examples: Use millimeters (mm) for mechanical engineering drawings,
kilopascals (kPa) for pressure,
megapascals (MPa) for stress, and
kilograms per cubic meter (kg/m^3) for density.

When convenient, choose prefixes resulting in numerical values between 0.1 and 1000, but only if this can be done without violating previously cited guidelines.

There are nine base and supplementary units plus many derived metric units formed from these. Of these units, mass, expressed in kilograms, is the only unit which has a prefix. It is important in technical calculations to be aware of the numerical magnitudes of the data being used to avoid erroneous results. To avoid error in calculations, prefixes should be converted to powers of 10, except when kilogram is involved; the kilogram is a base unit and it should be the unit used.

Example: 5 MJ = 5 x 10^6 J
4 Mg = 4 x 10^3 kg
3 Mm = 3 x 10^6 m

Avoid the use of prefixes in a denominator. The use of kilogram is excepted since it is a base unit.

Example: Write kJ/s, not J/ms
Write kJ/kg, not J/g
Write MJ/kg, not kJ/g

---

[5] The liter and the "are" would not normally be considered for use with another prefix except milli and hecto, respectively.

Do not use a mixture of prefixes unless the difference in size is extreme or a technical standard requires certain units.

Example: Use "The plywood was 44 mm wide and 1500 mm long," not "The plywood was 44 mm wide and 1.5 m long."

Use "I have a spool of 2-mm-diameter wire 1500 meters long."

Do not use multiple units or multiple prefixes.

Example: Use 15.26 m, not 15 m 260 mm. Use milligram (mg), not microkilogram (μkg)

Do not use a prefix without a unit. For example, do not use kilo for kilogram and do not use megs for megohms.

Symbols for Angles and Temperature

The symbol for angular degree (°) and degree Celsius (°C) should **always** be used in USBR publications and correspondence when giving a measurement. However, when describing the measurement scale and not a specific measurement, use the full name.

Examples: Angles will be measured in degrees and not radians. The angle of inclination was 27.6°.

Most temperatures will be given using the Celsius scale; the Kelvin scale will have limited application. The normal body temperature is 37 °C.

When giving a series of temperature values or a temperature range, use the measurement symbol after the last value only.

Examples: The winter daytime temperature in Denver ranges from −5 to 15 °C.

The test chamber readings for the low-temperature test cycle were 178, 192, 210, and 250 K.

It is technically correct to use SI prefixes with names and symbols such as degree Celsius (°C), kelvin (K), and angular degree (°). However, it is preferred that this practice be avoided; the resulting units, names, and symbols are awkward, difficult to recognize, and easily subject to misinterpretation. It is preferred that the numerical coefficient be adjusted so these units can be used without prefixes.

CHAPTER 2—SI STYLE AND EDITORIAL GUIDELINES

Symbol Modifiers

In the U.S. customary system, the practice of attaching modifiers to symbols and abbreviations to further quantify the unit of measure has developed. Examples of this include "psia" and "psig" to indicate pounds per square inch, absolute and gage, respectively. In the SI system, attachment of letters to a unit symbol as a means to further identify the measure is incorrect. See table 2-2 for examples of modified symbols and how they may be written.

Table 2-2.—Modified symbols

Customary units with modifiers and incorrect SI symbols	Express in the following manner for SI units
psig (gage pressure)	*gage pressure of 13 kPa or 13 kPa (gage)
psia (absolute pressure)	*absolute pressure of 13 kPa or 13 kPa (abs)
VAC, Vac (volts alternating current)	*V a.c. or V (a.c.)
MWe (megawatts electrical, power)	MW
kJt (kilojoules, thermal)	kJ

* Always maintain a space after the SI unit symbol and any modifying information.

As indicated by the last two entries in table 2-2, physical quantities such as energy and power are measured in a stated SI unit; there is no need to identify the source of the quantity since 100 watts equals 100 watts, regardless of whether one is electrical and the other is mechanical power.

A further example is: Engine ratings in brake horsepower (bhp) will be superseded by ratings given in kilowatts (kW).

Algebraic Equations and Formulas

In presenting equations or formulas in Bureau publications and technical documents, two conditions may arise regarding SI and U.S. customary units of measure. The first condition is that the original and most commonly used equation uses only U.S. customary units or a mixture of SI and U.S. customary units. When this occurs, it is better to restructure the equation or formula (if necessary)[6] and present it using only SI units of measure. The equation using only SI units should be the prominent equation presented. The equation using the U.S. customary units or the SI/U.S. customary mixture

---

[6] Many equations do not change when the units are converted to SI. Use dimensional analysis to determine the impact on any constants and multiplication factors.

may also be presented using an extra paragraph, a footnote, or a parenthetical insertion.[7] The conversion factor changing the resultant U.S. customary unit to the appropriate SI unit may be presented in a footnote if knowledge of the relationship will be helpful.

Secondly, equations which are already SI or dimensionless are to be presented in their original format. No reference should be made to U.S. customary units and no conversion factors are to be provided.

Presentation of Chemical Elements

The use of SI and the writing of chemical units are not directly related and it is most likely that in the majority of USBR publications where these two entities coexist, there will be no problem. However, examination of the periodic table does reveal several chemical symbols which duplicate SI symbols. Examples of this include C which represents coulombs in SI and carbon in the chemical table, and Mg which represents megagram in SI and magnesium in the chemical table.

All technical writers are asked to take the necessary care to prevent misinterpretation by the reader regarding the use of SI units and chemical symbols. To aid the writer, there are several guidelines which are important to remember; these are:

1. Never use an SI symbol or chemical symbol to start a sentence.

2. SI units are usually written in full if not being used with a numerical measure. There is always a space between the numerical coefficient and the SI symbol.

3. The numerical coefficient and the chemical symbol are written together without any space between. This is the principal difference in writing an SI value and a chemical unit.

The following is presented as a uniform guide for USBR engineers/scientists in presenting chemical units:

Mass number —When presenting the mass number, place it as a superscript to the left of the nuclide. Examples of this include ^{12}C and ^{14}N for carbon-12 and nitrogen-14, respectively.

Atomic number —The atomic number should be placed as a left subscript. Examples of this are $^{14}_{6}C$ for carbon-14, and $^{235}_{92}U$ for uranium-235.

Ionization number—Two styles of showing the ionization state are acceptable. The preferred method is to use a numerical right

---

[7] Do not attempt to show both SI and U.S. customary units in a single equation with the use of parentheses. An example of this incorrect procedure is: F, N (lbf) = M, kg (lbm) x a, m/s^2 (ft/s^2).

superscript. Examples are Ca^{+2} and SO_4^{-2}. The alternative method is to use an appropriate number of positive or negative signs; the previously used examples would then be Ca^{++} and SO_4^{--}, respectively. The disadvantage of the second method is that a large surplus (or shortage) of electrons results in a cumbersome symbol.

Neutrons — The number of neutrons in the nucleus is shown as a right subscript. For example, the isotope of calcium-40 containing 20 protons (its atomic number) and 20 neutrons in its nucleus is written $^{40}_{20}Ca^{20}$.

Excited states — Show excited states by adding to the left superscript. For example, ^{110m}Ag indicates an excited state of a silver-110 nucleus, and *He indicates an excited state of a helium atom.

Abbreviated Summary of SI Style Notes

Names of units:
1. The name of a unit begins in lowercase, except at the beginning of a sentence.
2. Apply only one prefix to a unit name. The prefix and unit name are joined without a hyphen or space between.
3. If a compound unit involving division is spelled out, the word "per" is used. Only one "per" is permitted in a written unit name.
4. If a compound unit involving multiplication is spelled out, the use of a hyphen is usually unnecessary, but can be used if clarity is aided.

Symbols:
1. Symbols are preferred when units are used in conjunction with numerals.
2. Symbols are never made plural.
3. A symbol is not followed by a period except at the end of a sentence.
4. Symbols for units named after individuals show the first letter capitalized. The symbol for liter (L) is also capitalized.
5. Symbols for all prefixes greater than kilo are capitalized; all others are lowercase.
6. Use numerical superscripts (^{2} and ^{3}) to indicate squares and cubes; do not use sq., cu., or c.
7. Exponents also apply to the prefix attached to a unit name; the multiple or submultiple unit is treated as a single entity.
8. Never start a sentence with a symbol.
9. Compound units formed by division contain a slash (/) to indicate division. Multiple slashes must be separated by parentheses.

10. Compound symbols formed by multiplication contain a product dot (·) to indicate multiplication.

11. Do not mix symbols and spelled out unit names.

Numerals:

1. A space is left between the last digit of a numeral and a symbol.
2. The period is used as a decimal marker.
3. A space is used instead of a comma to group numerals into 3-digit groups.
4. Decimal fractions are generally preferred to common fractions.
5. Decimal values less than one have a zero to the left of the decimal marker.
6. Multiples and submultiples are generally selected so that the numeral coefficient has a value between 0.1 and 1000. Similar quantities should use the same unit even if the values fall outside this range.
7. Do not substitute the product dot in place of a multiplication sign (x).
8. Use numeric digits, instead of spelled out words, for all unit coefficients.

Chapter III

FEATURED UNITS

Introduction

In this chapter, SI units and physical quantities which will be commonly used by the Bureau of Reclamation personnel are presented. The SI units are specifically discussed to enable a greater understanding of the magnitude of each measure and how the units are to be used.

The SI units preferred for use within the Bureau are identified in tables 3-5 through 3-14. Additional units which are not SI, but acceptable alternatives, are also listed along with any applicable limitations or restrictions.

To aid the reader in acquiring a physical and mental recognition of SI units, a section identifying common modes of comparison is included. The unit comparisons are limited to particular nontechnical measures commonly used.

Units and Quantities

Area

Area is being discussed to introduce the hectare (ha); the hectare is a special name for the square hectometer (hm^2). The hectare equals $10\ 000\ m^2$ and is used in the same manner as the U.S. customary acre.

Replacing the acre, the hectare will be used as the preferred measure of land and water areas. However, the square meter (m^2) remains the preferred SI unit for other measures of area.

The hectare is equal to approximately 2.5 acres. Other physical examples of a hectare include a football field including end zones (fig. 3-1) is approximately 0.5 ha, and the whole playing area (fair and foul) of a major league baseball field is approximately 1 ha.

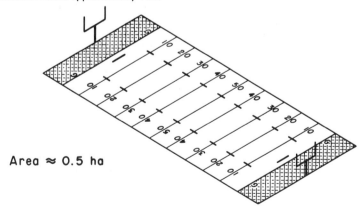

Figure 3-1.—American football field—0.535 hectare. 40-D-6335

Angles

The SI unit for plane angles is the radian; there are 2π radians in a complete circle. Within the Bureau of Reclamation, the use of the radian will be limited to scientific calculations. The degree will continue to be the preferred unit for angular measurement.

For drawings and specifications, the angles shall be written in degrees-minutes-seconds or degrees and decimal fractions (three decimal positions), as required. This format will be used in publications also, with the decimal format being preferred; show only the required number of decimal places up to a maximum of three.

The revolution per minute (r/min) will remain the preferred measure for describing rotating machinery. The conversion to radians per second (rad/s) will only be necessary in calculating forces, stress, inertia, torque, and other physical quantities which will be measured in SI units. Noting the equivalent radian per second (rad/s) value in parentheses with the revolution per minute (r/min) values is optional.

Common Fractions

In written material, the use of common fractions (i.e., 1/2, 1/3, 1/4, etc.) is an acceptable means of expressing an imprecise measure or a nominal hardware/material size. Examples of an imprecise measure would be 1/2 day, 1/4 mile, or "1/3 of the time." For measures such as these, it would be illogical to use three-place decimal numbers which would indicate a precision not intended.

On drawings and in written material, it is preferred that decimal numbers be used instead of noncommon fractions. This applies to U.S. customary dimensions and especially to SI dimensions; avoid the use of fractions with SI units on drawings or in written material. In specifications, the Bureau of Reclamation will continue to purchase some materials built to U.S. customary units; items such as conduit, tubing, and fasteners will be described by their nominal dimension (i.e., 3/4-inch conduit, 1/4-inch tubing, 1/2-13-UNC cap screw). When it is common practice to use fractions for material callouts, continue to do so until the industry changes its standard designations or metric materials become available.

Energy

The derived unit for energy (or work) is the joule (J). The joule is the special name given to 1 newton meter of energy; 1 N·m is the work done when the point of application of 1 N of force is displaced through a distance of 1 m in the direction of the force.

Note that the unit for newton meter (N·m) (torque and moment) and the definition unit for the joule (1 N·m) are dimensionally equal. The two concepts are quite different when considered vectorially. Note that in figure 3-2, the two concepts are depicted with vector representations; the different orientations of force and length are quite obvious.

CHAPTER 3—FEATURED UNITS

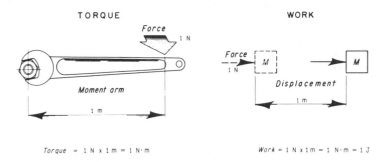

Figure 3-2.—Vector comparison of work and torque. 40-D-6336

Since vector representation is seldom employed when expressing numerical quantities, it is recommended that the joule be used when presenting energy values and the newton meter be used exclusively for torque and moment. Expressing torque in newton meter per radian (N·m/rad) clarifies its relationship to energy since the product of torque and angular rotation is energy.

$$(N \cdot m/rad) \, rad = N \cdot m = 1 \, J$$

Never use the joule as a measure for torque or moment.

There are many forms of energy; these include mechanical, electrical, thermal, kinetic, and several others. Eventually, it is expected that most, if not all, quantities of energy will be expressed in joules. However, units such as the electronvolt (eV) and the kilowatt-hour (kW·h) will remain in use for some time in their respective fields, nuclear physics and the electrical power industry; it is not unreasonable to assume they may also be replaced by the joule. Units such as calorie, horsepower-hour, and Btu will be phased out much more rapidly.

Force, Weight, and Mass

In SI, there is one basic unit for force and one basic unit for mass; these units are the newton (N) and the kilogram (kg), respectively. In the U.S. customary, Imperial, and traditional cgs-MKS metric systems, the units to be used for force and mass are many and are confusing, at best. And there is always the question concerning weight; is this a mass or is it a force, and will the same definition be applicable everywhere, including other parts of the solar system?

Listed in table 3-1 are the comparable units of measure for SI and the other systems of measure.

In the U.S. customary system,[1] the pound has played a dual role; it is the unit most people use to describe both mass and force. This dual relationship was perpetuated in the traditional metric systems with the use of kilogram and

---

[1] The U.S. customary system is defined as the inch-pound system of measurement.

Table 3-1.—*Elements of mass, force, and force of gravity*

Physical quantity	SI	Other systems
Gravitational acceleration (g)	9.806 65 m/s^2	32.174 05 ft/s^2 980.665 cm/s^2
Mass	kilogram	slug pound (mass)
Force	newton	poundal pound (force) pound (weight) kilogram (force) kilopond

kilogram (force); this dual identity is not permitted in SI. The customary term of weight is not used in SI; it also carries a double definition.

To clarify the difference between SI and customary units, the following definitions and discussions are presented so that these units of measure can be fully understood and an accurate comparison with the SI units can be made.

Weight.—Weight has been the method of describing an object's mass by measuring how much gravitational force the object exerts. This relationship has been widely accepted even though it is not constant because gravitational forces vary throughout the world and are significantly different in outer space. On Earth, a 200-pound man "weighs" 200 pounds, but on the Moon, he "weighs" only 33 pounds.

Although weight is more commonly used to describe mass, it is actually a measure of force, the force resulting from gravity.

Pound (weight) and pound (force).—The pound is the traditional unit of measure for the weight of an item. Since an object's weight will change with the gravitational field and since mass is constant, it follows that pound is a unit of force. The term "pounds (force)" is commonly used and widely accepted; however, identification as a force is redundant and not necessary.

Pound (mass).—The use of this term is a means of expressing mass as a function of the Earth's gravity; it is a quantity (mass) which will exert the force of 1 pound at standard gravitational attraction. The use of "pound" alone is technically a force; in common use, it is generally and incorrectly used as a mass term.

Slug.—The slug is technically the U.S. customary unit of measure for mass. This terminology has very limited recognition, which contributes to the difficulty in differentiating between weight (force) and mass.

A slug subjected to a force of 1 pound will move with an acceleration of 1 ft/s^2, and a slug subjected to gravitational acceleration exerts a force of 32.174 pounds. Note that Newton's law states: force equals mass times acceleration; $F = m \times a$.

CHAPTER 3—FEATURED UNITS

Poundal.—The poundal is a unit of force; it is the force, which acting on a mass of 1 pound (0.031 slug), gives it an acceleration of 1 foot per second squared. The pound equals 32.174 poundals; note that the conversion factor equals gravitational acceleration (g). So as gravity varies, so does the relationship between the pound and the poundal.

Kilogram (force).—This terminology is a means of expressing the force exerted by a mass of 1 kg at mean sea level; this equals 9.806 65 newtons. This terminology is not SI and its use is an erroneous attempt to perpetuate the confusion of the U.S. customary system regarding weight, force, and mass. Do not use this terminology.

The slug and the poundal represent attempts to remove gravity as a factor and introduce a degree of coherency into the customary system. The failure of these two units to replace one of the pound's two definitions can be related to the fact that they were really not needed as long as man was relatively isolated and restricted to Earth; standards of trade were not a problem and there was no compelling need to consistently differentiate mass from force; they were considered one and the same. The kilogram (force) which was derived in the traditional metric system is the "metric version of weight and mass."

It is very easy to detect the source of confusion using the U.S. customary system of weight, force, and mass. The SI units described in the following do not have the dual definition problems which exist in the customary system.

Kilogram.—The kilogram is a measure of an object's mass; the mass of an item is constant and does not change with the gravitational field or its rate of acceleration. The use of the term "kilogram (mass)" is redundant and should be avoided.

Newton.—The newton is the SI unit of force. Using Newton's equation of $F = m \times a$, a newton is 1 kilogram times 1 meter per second squared. Acting under Earth's gravitational pull, a mass of 1 kg exerts a force of **9.806 65** newtons.

Specifically, it would be better to use "little mass" or "small mass" or "low density," and "much mass," "large mass," or "high density" to describe objects currently described as light or heavy, respectively. The use of terminology such as "weighted average" shall continue unchanged.

A general summary of SI rules covering mass and force is listed below:
1. The SI unit for mass is the kilogram.
2. The SI unit for force is the newton. Do not use kilogram (force).
3. Gravity is not an essential element of the SI system.
4. The use of weigh and weight should be avoided; they should not be used to indicate the measuring process or the measure of mass. Recommendation: Rather than say "it weighs," say "it has a mass of." Rather than "weighing the object," say "measuring the object's mass." Similar sentence structure changes can be made for other tenses, subjects, etc.
5. The use of light and heavy is still acceptable terminology, even if weight is part of the description (i.e., lightweight).

When using SI units, the "g" factor is not used in dynamics problems, but must be introduced in problems relating to statics. This is the reverse of procedures used when the customary gravitational system is involved.

A possible alternative to weight is to use the force of gravity. With such a description, there would be a clear distinction between the mass of an item and the force it exerts under gravitational attraction.

With the SI units for force and mass, some people may be concerned as to what units metric scales will display, and what type of scales will be used. Both units of measure will be used and the traditional spring scale will be the most common measurement device. The spring scale can either be calibrated to measure mass as a function of its gravitational force, or it may provide a direct reading in newtons because of the gravitational force. A scale calibrated in kilograms will have an accuracy of plus and minus 0.5 percent anyplace on Earth, but will be grossly incorrect if used on the Moon. A scale calibrated in newtons could be used anywhere and would maintain an accuracy determined by its calibration.

The best type of scale for the measurement of masses is the balance beam scale. Neglecting the negligible buoyancy effects of air, the balance beam scale will provide highly accurate values.

Linear Measure

As previously stated, the meter is the fundamental SI unit for length or distance. The Bureau of Reclamation also will be using the kilometer, the millimeter, and less frequently, the micrometer as units of length. These multiples and submultiples are in accordance with the recommended practice of selecting prefixes which are multiples of 1000. The use of the millimeter will predominate since it is the preferred unit for most engineering measures and other technical quantities.

The centimeter, often referred to as the "metric inch," will find most of its application with consumer products and nontechnical measures. Technical use of the centimeter will be very limited; quantities such as water quality (specific conductivity, $\mu S/cm$) and snow depths will use the centimeter. The technical use of the centimeter is being discouraged primarily because it does not fit into the multiples series based upon 1000.

In technical documents, when writing values given in meters or kilometers, it is recommended that no more than three decimal positions be used. Values expressed in millimeters should be whole numbers unless necessary precision requires otherwise. Use micrometer (μm) or nanometer (nm) for more exacting numbers required in physics and chemistry. Judicious application of these guidelines and maintenance of the prefix selection based upon multiples of 1000 will virtually eliminate any possible error in interpretation of a quantity value.

The Liter

Dimensions.—The liter is a volumetric measure originally introduced in 1795; it was intended to equal the cubic decimeter. This definition was changed in 1901 by the 3rd CGPM. The conference decided that the liter's dimensions would be determined by the volume occupied by 1 kilogram of pure water at its maximum density under normal atmospheric pressure. This

CHAPTER 3—FEATURED UNITS 35

volume was determined to be equal to 1.000 028 dm^3, very close to the cubic decimeter but not exactly equal.

In 1964, the 12th CGPM reversed its previous decision and established the liter as a special name for the cubic decimeter. It also recommended that the liter should not be employed to give the results of high accuracy volume measurements; this is because of the two definitions of the liter which have been used. The liter is permitted in SI but its use by the Bureau of Reclamation should be restricted to measures of liquid and gas.

Prefixes.—Liter has only one preferred and commonly recognized prefix— "milli" (m). Several other countries have adopted the use of kiloliter and megaliter to describe large volumes of liquid; this usage has been taken into consideration by including appropriate conversion factors in the chapter V volume conversion table. Also, the microliter (μL) is used in some areas of research. However, Reclamation personnel are not to use any prefix other than milli; large volumes will be expressed in cubic meters (m^3), cubic dekameters (dam^3), cubic hectometers (hm^3), or cubic kilometers (km^3).

Symbol.—The symbol for liter has been the center of many debates. The reason for this is that on most typewriters there is no difference between lowercase "el" (l, the ISO designated symbol for liter) and the figure "one" (1).

The script "el" (ℓ) has been used extensively to avoid confusion with the number one, but not all typewriters have this symbol. Also, this symbol has not gained international acceptance as an alternative to the Roman "el".

The Bureau of Reclamation shall use the capital L as the symbol for liter. This is in accord with the decision of NBS (National Bureau of Standards), ASTM (American Society for Testing and Materials, and ANMC (American National Metric Council). The recommendation to CIPM (International Committee for Weights and Measures) to revise the SI symbol for liter has been rejected by that committee. The international symbol for liter remains the lowercase Roman l; however, additional efforts to make L the new international symbol for liter are expected. The 1975 Metric Act provides the authority for the Secretary of Commerce to implement this deviation from the official SI guidelines.

The Metric Ton

Three different names for the metric ton currently exist, all of which are used to describe the same amount of mass, 1000 kg. These names are megagram (Mg), tonne (t), and metric ton (t).

Arguments can be made for and against each of these names. Some of these arguments are:

1. The megagram is SI; metric ton and tonne are not.
2. The symbol for tonne and metric ton is t; this may be mistaken to be the U.S. customary short ton. The symbol for megagram (Mg) is distinct and avoids all confusion.
3. The megagram is not widely recognized and is not readily identified by nontechnical personnel as being the equivalent of a "metric ton." Due to the publicity surrounding recent international grain sales, the general public is very familiar with the metric ton, and to a lesser extent with tonne.

These arguments are presented for information only; all three terms probably will be encountered. The preferred terms are megagram for technical measures and the metric ton for commercial measures. Although the use of tonne is very common in Canada, Australia, and European countries, its use is not advocated within the United States or by the USBR.

Pressure and Stress

Pressure.—A pressure is a force per unit area; the SI unit of force is the newton (N) and the base unit of area is the square meter (m^2). In SI, the expression, newton per square meter (N/m^2), is given the special name pascal (Pa).

The pascal is a very small pressure; a physical example of this pressure can be illustrated by pouring 100 mL of water over an area of 1 m^2. This will result in water 0.1 mm deep, and this equals the pressure of 1 Pa. A dollar bill lying flat on a surface also exerts a pressure of 1 Pa.

Since the pascal is so small, all pressure should be given in kilopascals (kPa). Strict adherence to the use of kilopascals may result in a violation of the numerical coefficient rule; this will be acceptable. The use of other mutiples of the pascal is not deprecated; the circumstance and judgment should determine what other unit is appropriate. Examples of common pressures expressed in SI are given in table 3-2.

Other measures of pressure will continue in use by the Bureau of Reclamation even though their use is discouraged. In meteorology, the bar and millibar (mbar) are acceptable on a temporary basis. Hydraulics engineers wish to continue to use pressure head expressed as a column of water; kilopascal is the preferred unit of measure, but meters of water and meters-head will be acceptable alternatives.

Stress and modulus.—The pascal is also used to express stress levels and modulus of elasticity values for materials. Because of the size of these values, stress values should be given in megapascals (MPa) and the modulus values should be in gigapascals (GPa). The use of gigapascals for modulus, megapascals for stress, and kilopascals for pressure provides a quick indication of the physical quantity being discussed.

Table 3-2.—*Common pressures expressed in SI*

auto tires	150-200 kPa
racing bike tires	520-570 kPa
hydraulic lines	10 000-18 000 kPa
1 mm of mercury	0.133 kPa
atmospheric pressure	95-105 kPa
100-ft head of water	300 kPa
pressure cooker	90-110 kPa
basketball/football	70-80 kPa

Relative Density

The term "specific gravity" has been the accepted dimensionless value describing the density ratio of solid materials relative to water (at 4 °C) and gases relative to air (at STP). The API (American Petroleum Industry) has proposed that specific gravity be abandoned in favor of relative density, a more descriptive term. The API proposal has been generally adopted as part of SI since it does disregard the reference to gravity.

The term "relative density" has traditionally been used in the earth sciences/soil mechanics area for a measure of soil compactness relative to a particular standard. Because this term is commonly used in the earth sciences, rather than risk the possible misinterpretations which could result from the use of two definitions for the same term, the Bureau of Reclamation has modified the terminology slightly to aid in the distinction of the two measures.

In the Bureau's work, specific gravity is to be replaced with "relative mass density." This terminology is not to be abbreviated or otherwise shortened. It is expected that as SI terminology becomes more widely adopted, the soil compactness ratio will acquire a new name; the use of "density index" has been proposed for this ASTM test. Several years after this occurs, relative mass density could be shortened to relative density; within the Bureau, a time lag would be absolutely necessary to preclude any lingering double definition problems. The continuing use of relative mass density would not cause any problems.

It will always be acceptable to describe materials in kilograms per cubic meter; the density value eliminates all confusion.

The Steradian

The steradian (sr) is the SI unit for a solid angle. Most people will have little need for knowledge of the steradian as it is a measure few will have occasion to use. However, this section is being included for those who may be interested in an intuitive concept of the steradian measure.

In the U.S. customary system, the concept which has been used to describe solid angles is the sphere; solid angles have been expressed as a portion of a sphere. A common example of a spherical sector is an ice cream cone; the top portion of the ice cream cone is a spherical segment.

In figure 3-3, a sector taken from a sphere is shown. This figure shall be used to develop the equations for determining the number of steradians when the plane angle a is known and for finding angle a for a given steradian measure.

When a solid angle taken from a sphere is bisected along the centerline into two equal parts, the plane angle at the vertex is equal to a. Half of this plane angle ($a/2$) is used in the equations which follow; these equations develop the relationship between a and ψ (the solid angle of the spherical sector). Listed are some of the dimensions shown on this conical sector of a sphere:

R = radius of sphere
ψ = solid angle of the spherical sector, measured in steradians
c = arc length along the spherical surface subtending the angle θ
a = the vertex plane angle when the cone is projected onto the x-y plane
R_1 = radius subtended by arc length c, and the radius subtending the angle θ
θ = plane angle, measured from the sector centerline, approaches $a/2$ as a limit
$\Delta\theta$ = incremental change in θ
Δc = incremental change in c resulting from $\Delta\theta$
ΔA = small area on top surface of shape

From figure 3-3, it can be seen that ΔA is a function of $\Delta\theta$ and R_1. The following relationships can be derived:

$$\Delta A = (\Delta c)(2\pi R_1)$$
$$\Delta c = R \Delta\theta$$
$$R_1 = R \sin\theta$$
$$\Delta A = (R\Delta\theta)(2\pi)(R \sin\theta)$$

Solving for A as θ approaches $a/2$ as a limit:

$$A = 2\pi R^2 \int_0^{a/2} \sin\theta \, d\theta$$

$$= 2\pi R^2 \left[-\cos\theta\right] \Big|_0^{a/2}$$

$$= 2\pi R^2 (1 - \cos a/2)$$

For unit circles ($R = 1$), the arc is equal to the angle (measured in radians) times the radius (1 mm, 1 m, 1 unit, etc.). A similar relationship applies to solid angles (steradians), the surface area for a unit sphere equals the radius squared times the solid angle (steradians). The surface area for a sphere is known to be $4\pi R^2$ and the value in steradians for a unit sphere is 4π sr.

Referring to the last equation, it is known that to make a complete sphere, $a/2$ must equal π. Inserting π into this equation, the area is found to equal $4\pi R^2$. For a unit sphere, R^2 equals 1 and is therefore disregarded.

For a unit sphere, the surface area (A) equals the number of steradians. Setting $\psi = A$, equations for determining angle a given ψ and determining ψ given a can be derived. These are:

$$a(\text{degrees}) = 2 \cos^{-1}\left[\frac{\psi - 2\pi}{-2\pi}\right]$$

$$\psi(\text{sr}) = 2\pi [1 - \cos a/2]$$

CHAPTER 3—FEATURED UNITS

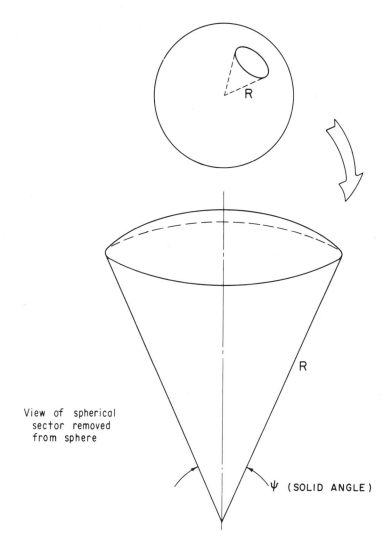

Figure 3-3.—Element analysis of spherical sector. Sheet 1 of 2. 40-D-6337-1

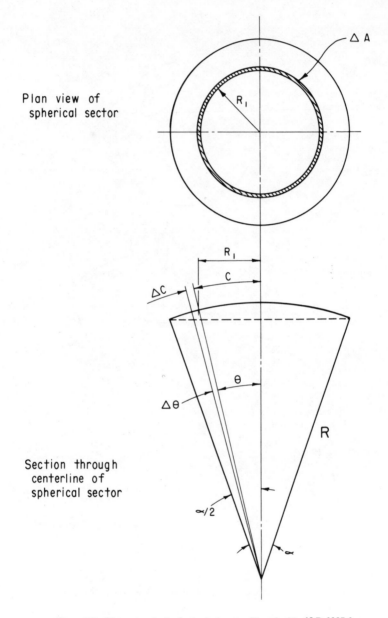

Figure 3-3.—Element analysis of spherical sector. Sheet 2 of 2. 40-D-6337-2

CHAPTER 3—FEATURED UNITS

These equations are true for spheres of all radii as the dependent variables are angles, either plane or solid. The spherical surface area of a spherical sector is computed by multiplying the solid angle (in steradians) times the square of the radius.

See table 3-3 for some example solid angle-plane angle relationships.

Table 3-3.—*Solid angle (ψ in steradians)—plane angle (a in degrees) corresponding values*

a	ψ	ψ	a
30°	0.214	0.5	46.024°
45°	0.478	1.0	65.541°
60°	0.842	1.5	80.848°
75°	1.298	2.0	94.048°
90°	1.840	2.5	105.957°

Figure 3-4 provides a graphic representation of the relationship between the solid angle (ψ) and the maximum included plane angle (a).

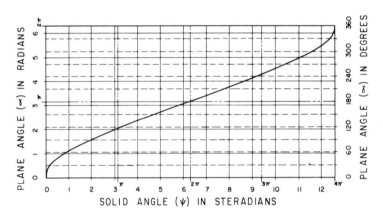

Figure 3-4.—Relationship of solid angle to the maximum included plane angle. 40-D-6360

Temperature

The kelvin (K) is the SI unit of thermodynamic temperature. The use of kelvin is generally limited to scientific calculations; degree Celsius (°C) is used as the common measure of temperature. The Celsius temperature (t_c) is related to kelvin temperature (t_k) by the equation: $t_c = t_k - 273.15$. The numerical constant in this equation, 273.15, represents the triple point of water 273.16 minus 0.01.[2] Temperature intervals are equal for both scales; 1 °C equals 1 K, exactly. Temperature intervals are expressed in the same manner as a temperature, so it is possible to say 40 °C equals 40 K; if this is done, it is essential that the text narrative indicate when temperature intervals are being presented.

What happened to centigrade? In 1948, the CIPM selected from three names proposed to denote temperature: (1) degree centigrade, (2) centesimal degree, and (3) degree Celsius. They selected degree Celsius and this was adopted by the 9th CGPM. One reason the use of centigrade was not continued is that some countries have used the grade (or gon) as an angular measure (i.e., a right angle equals 100 grades); to them, a centigrade equals 0.01 percent of a right angle. As we will continue to use angular degrees, European countries will continue to use grade, therefore, the possible confusion existed between centigrade (0.01 grade) and centigrade (a temperature).

The symbol for degree Celsius is °C. The capitalization of degree Celsius is often questioned as violating the SI style for units named after individuals. The rationale for this style is that degree is the measure with Celsius being the modifier.

Anders Celsius, 1744, a Swedish astronomer, developed the centigrade scale; the scale now bears his name in his honor.

The new Celsius scale is essentially equal to the old centigrade scale; there was a slight change in 1960 when the General Conference on Weights and Measures established two new reference points. The new reference points are absolute zero and the triple point of water; these replace the freezing and boiling points of water, respectively. This change is only significant to certain experts and does not alter the close approximation of water freezing at 0 °C and boiling at 100 °C.

Time

Intervals.—In SI, the base unit for time is the second (s). This unit of time is preferred and should be used wherever practicable; technical calculations almost always require use of the second when expressing rates.

Other units of time will continue to be used; namely, calendar cycles such as year and day, and particularly multiples of the second, such as minute and hour. Examples of these include automobile velocity in kilometers per hour (km/h) and machinery rotational speed in revolutions per minute (r/min).

Calendar cycles such as day, week, month, and year should be avoided as there are several interpretations which may be applied to each. If used, the calendar period should be defined; for example, is a month 1/12 of a calendar year or is it a nominal 30 days?

Calendar dating.—A new sequence of digits has been proposed by the ISO for numeric dating. Numeric dating is the expression of dates by means of

---

[2] The zero point of the Celsius scale has an offset of 0.01 from the Kelvin scale; this means the triple point of water occurs at 0.01 °C.

numbers. In this new sequence, the numbers give the year first followed by the month, and then the day. For example, June 20, 1945, this would be written 1945-06-20.

This system of dating shows the logical sequence in which data are chronologically stored. It can also be extended to include the time of day, given in four digits, based on the 24-hour clock.

This system of dating **has not** been adopted for use by the Bureau of Reclamation; future adoption is not precluded.

Twenty-four-hour clock.—The use of the 24-hour clock eliminates the use of a.m. and p.m. Giving the time from midnight to 12:59 remains unchanged; the remaining customary afternoon times are increased by 12.

The Bureau of Reclamation **is not** adopting the use of the 24-hour clock.

Torque

Torque is the vector product of force and moment arm. Force is in newtons and the moment arm is in meters; therefore, torque is expressed in newton meters (N·m).

Common torques expressed in SI are listed in table 3-4.

Table 3-4.—*Common torque values*

Seating spark plugs	40 N·m
Cylinder head bolts	130 N·m
Turning a wood screw	3 N·m
A 4-cylinder auto engine output	160 N·m
A V-8 auto engine output	550 N·m

As previously discussed, the newton meter is also used to define energy; however, the newton meter is given the special name joule (J) for describing energy. Do not use the joule as a measure of torque.

USBR Preferred Units of Measure

The following tables list the SI units of measure selected for use within the Bureau of Reclamation. The units are listed by engineering and scientific fields and are collated with a particular physical quantity within that field. Units were selected based upon ISO 1000 [30], the Canadian Metric Practice Guide [12], ASTM E-380 [32], and the USBR Interim Metric Guide [31]. Selections were made in accordance with the general recommendations of the E&R Center metric units subcommittee. Deviations from the accepted SI units are allowed in several cases where it is necessary to accommodate universally accepted units for particular fields of engineering and science.

In many instances, several alternative units (multiples or submultiples) are listed; *to indicate the truly preferred or normally used unit, a left superscript asterisk is shown*. Limitations or restrictions to the use of particular units are listed in the "remarks" column.

The following is a list of the preferred units tables:

3-5—USBR preferred units—Space and time
3-6—USBR preferred units—Periodic and related phenomena
3-7—USBR preferred units—Mechanics
3-8—USBR preferred units—Heat
3-9—USBR preferred units—Acoustics
3-10—USBR preferred units—Electricity and magnetism
3-11—USBR preferred units—Light and electromagnetic radiations
3-12—USBR preferred units—Physical chemistry and molecular physics
3-13—USBR preferred units—Nuclear physics and ionizing radiations
3-14—USBR preferred units—Water resources engineering

Table 3-5.—*USBR preferred units—Space and time*

Quantity	Preferred units—symbol(s) and/or name(s)	Remarks
Acceleration—		
angular	rad/s² (radian per second squared) mrad/s²	
linear	m/s² (meter per second squared)	
Angles—		
plane	degree (°), minute ('), second (") *degree and decimal fraction rad (radian)	Restrict the use of radians to scientific and engineering calculations.
solid	sr (steradian)	
Area—		
tracts of land and water	ha (hectare)	Do not use "are" by itself or with a prefix other than hecto.
very large areas	km² (square kilometer)	
other	m² mm² cm²	The technical use of square centimeters should be restricted to particular uses adopted by technical societies or engineering disciplines; it is generally to be avoided.
Length—		
stationing	m (meter)	One station equals 100 m.
elevations	m	
design drawings	mm m	Use meters for all large-scale civil drawings, see table 4-13 for examples.

Table 3-5.—*USBR preferred units—Space and time*—Continued

Quantity	Preferred units—symbol(s) and/or name(s)	Remarks
Length—Continued		
large distances	km	
consumer products	cm (centimeter)	
precipitation	mm (millimeter)	Snow is measured in centimeters.
reservoir depth	m	The use of the nautical mile (nmi) is acceptable in certain fields.
other	km *m mm μm (micrometer) nm (nanometer)	
Time—		
nominal engineering/ scientific units	*s (second) ms μs ns	
extended periods	a (year) d (day) h (hour) *min (minute)	Avoid the use of week and month as a time base. When a calendar period is used, define it.
Velocity—		
angular	r/min (revolution per minute) rad/s (radian per second)	Limit the use of radian per second to scientific and engineering equations.
linear	*m/s km/h	The use of knots is acceptable in certain fields.

Table 3-5.—*USBR preferred units—Space and time*—Continued

Quantity	Preferred units—symbol(s) and/or name(s)	Remarks
Volume—		
reservoir capacity and other large volumes	m^3	The USSR uses cubic kilometers (km^3). Australia uses megaliter (ML) and kiloliter (kL). The Bureau has decided to discontinue exclusively using cubic dekameter (dam^3). Large volumes expressed in cubic meters (m^3) may use scientific notation. The use of other SI volume units (e.g., hm^3, dam^3, etc.) is permissible.
small volume	m^3 *L (liter) mL (milliliter)	It has been recommended that the liter and its multiples not be used for high-precision measurements. This limitation will not apply in Bureau work.
gasoline engine sizes	*cm^3 L	

* Indicates unit with greater preference.

Table 3-6.—*USBR preferred units—Periodic and related phenomena*

Quantity	Preferred units—symbol(s) and/or name(s)	Remarks
Frequency	mHz *Hz (hertz) kHz GHz	
Periodic time	min *s ms μs ns	
Rotational frequency	*min^{-1} (per minute) s^{-1} (per second)	Rotating machinery is most commonly specified in revolutions per minute. Do not use radian per minute (rad/min); minimize use of revolution per second (r/s).
Wavelength	m *mm μm nm pm (picometer)	
Wave number	m^{-1} (per meter)	

* Indicates unit with greater preference.

CHAPTER 3—FEATURED UNITS

Table 3-7.—*USBR preferred units—Mechanics*

Quantity	Preferred units—symbol(s) and/or name(s)	Remarks
Compressibility	Pa^{-1} (reciprocal pascal)	
Density—		
area	*kg/m^2 (kilogram per square meter) g/m^2	
linear	*kg/m g/m mg/m	
mass	*kg/m^3 Mg/m^3 t/m^3 (metric ton per cubic meter) g/L *mg/L g/cm^3	
Energy/work	J (joule) MJ GJ	Joule is the preferred unit, but watt-hour (W·h) and its multiples may be used as an alternative in the field of electric energy consumption. The use of electronvolt (eV) is permitted in the field of nuclear physics.
Flow rate—		
mass volume	kg/s m^3/s L/s L/min	
Force	N (newton)	Never use kilogram force (kgf) or kilopond.
Fuel consumption, automotive	L/100 km	To convert from miles per gallon, divide **235.215** by the mile per gallon value.

Table 3-7.—*USBR preferred units—Mechanics*—Continued

Quantity	Preferred units symbol(s) and/or name(s)	Remarks
Inertia— area line mass	 m^4 mm^4 m^3 (meter cubed) mm^3 $kg \cdot m^2$ (kilogram meter squared)	
Mass	*kg Mg t (metric ton)	 The use of tonne is not approved.
Modulus of elasticity (Young's), shear rigidity, or bulk compression	*GPa MPa kPa	It is preferred that all modulus values be given in gigapascals even if the numerical coefficient rule must be violated. The use of other units must be determined by the particular situation.
Moments— force inertia (mass) momentum second moment of area	 $N \cdot m$ (newton meter) $kg \cdot m^2$ $kg \cdot m^2/s$ m^4 mm^4	
Momentum— angular linear	 $kg \cdot m^2/s$ $kg \cdot m/s$	
Power	W (watt) kW MW GW	Power is the rate at which work is done. One watt of power equals 1 joule per second (1 J/s).

CHAPTER 3—FEATURED UNITS 51

Table 3-7.—*USBR preferred units—Mechanics*—Continued

Quantity	Preferred units symbol(s) and/or name(s)	Remarks
Pressure	Pa *kPa MPa	Hydraulic head may be expressed in meters-head or meters of water. It is preferred that all pressures be given in kilopascals even if the numerical coefficient rule must be violated. The use of other units must be determined by the particular situation.
	bar (bar) mbar (millibar)	Restrict the use of bar and mbar to the field of meteorology.
Section modulus	m^3 (meter cubed) mm^3	Also line moment of inertia.
Stress	kPa *MPa GPa	It is preferred that all stress values be given in megapascals even if the numerical coefficient rule must be violated. The use of other units must be determined by the particular situation.
Surface tension	N/m	
Torque	N·m	Never use the joule as a measure of torque.
Viscosity— dynamic kinematic	 Pa·s (pascal second) m^2/s	 The rhe is given in "per pascal second."

* Indicates unit with greater preference.

Table 3-8.—*USBR preferred units—Heat*

Quantity	Preferred units—symbol(s) and/or name(s)	Remarks
Coefficient of heat transfer	$W/(m^2 \cdot K)$ *$W/(m^2 \cdot °C)$	
Enthalpy	J (joule)	
Entropy	J/K (joule per kelvin)	
Heat capacity	J/K kJ/K	
Heat flow	W (watt) kW	
Heat quantity	mJ J kJ MJ GJ	
Internal energy	J	
Linear expansion coefficient	K^{-1} (reciprocal kelvin) *$°C^{-1}$	
Radiant intensity	W/sr (watt per steradian)	
Specific enthalpy	J/kg	
Specific entropy	$J/(kg \cdot K)$ $kJ/(kg \cdot K)$	
Specific heat capacity	$J/(kg \cdot K)$	
Specific internal energy	J/kg	
Specific latent heat	J/kg kJ/kg MJ/kg	
Temperature— thermodynamic common measure	K (kelvin) °C (degree Celsius)	
Temperature interval	*°C K	
Thermal conductivity	$W/m \cdot K$ $W/(m \cdot °C)$	
Thermal resistance— mechanical	$m^2 \cdot K/W$ *$m^2 \cdot °C/W$	Commonly known as the R value. ISO proposing "thermal insulance" for this term.

Table 3-8.—*USBR preferred units—Heat*—Continued

Quantity	Preferred units—symbol(s) and/or name(s)	Remarks
Thermal resistance—Continued		
electronic	K/W *°C/W	
Thermal resistivity	m·K/W *m·°C/W	

* Indicates unit with greater preference.

Table 3-9.—*USBR preferred units—Acoustics*

Quantity	Preferred units—symbol(s) and/or name(s)	Remarks
Acoustic impedance	Pa·s/m^3	
Density (mass)	kg/m^3	
Equivalent absorption	m^2	
Frequency	Hz kHz MHz	
Mechanical impedance	N·s/m	
Period (time)	*s ms μs	
Reverberation time	s ms	
Sound particle velocity (inst.)	m/s mm/s	
Sound energy flux—sound power	W kW mW μW	
Sound intensity	W/m^2 mW/m^2 μW/m^2	
Sound power level	dB (decibel)	Reference power level must be stated.
Sound pressure (inst.)	Pa mPa μPa	
Sound pressure level	dB	Reference pressure level must be stated.
Specific acoustic impedance	Pa·s/m	
Static pressure	Pa mPa μPa	
Velocity of sound	m/s	
Volume velocity (inst.)	m^3/s	
Wavelength	m mm	

* Indicates unit with greater preference.

CHAPTER 3—FEATURED UNITS

Table 3-10.—*USBR preferred units—Electricity and magnetism*

Quantity	Preferred units—symbol(s) and/or name(s)	Remarks
Active power	GW MW kW W mW µW nW	In the electric power field, "apparent power" may be expressed in volt amperes (V·A) and "reactive power" is expressed in volt ampere reactive [V·A (reactive)]. The IEEE has recommended continued use of the var even though it violates SI symbol guidelines.
Admittance (susceptance, conductance)	kS S (siemens) mS µS	1 mho = Ω^{-1} = 1 S
Capacitance	F (farad) µF nF pF	For reciprocal farad, use 1/F or F^{-1}; do not use the daraf.
Charge density— volume surface	 MC/m^3 kC/m^3 C/m^3 (coulomb per cubic meter) mC/m^3 MC/m^2 kC/m^2 C/m^2 mC/m^2	
Conductance	kS S (siemens) mS µS	
Conductivity	S/m kS/m MS/m	In limnology, water chemistry, and oceanographic literature, the use of microsiemens per centimeter (µS/cm) is the accepted unit for specific conductance.

Table 3-10.— *USBR preferred units—Electricity and magnetism*—Continued

Quantity	Preferred units—symbol(s) and/or name(s)	Remarks
Current density—		
area	A/m^2 (ampere per square meter) kA/m^2	
linear	A/m kA/m	
Displacement	C/m^2	
Electric charge	kC C (coulomb) mC μC	1 ampere-hour (A·h) = 3600 coulombs
Electric current	A (ampere) mA μA	
Electric dipole moment	C·m	
Electric energy	J (joule) kJ MJ	
Electric field strength	mV/m V/m (volt per meter) kV/m	
Electric flux	MC kC C mC	
Electric polarization	kC/m^2 C/m^2 mC/m^2	
Electric potential	kV	
Electromotive force	V (volt) mV	
Impedance, resistance, or reactance	MΩ kΩ Ω (ohm) mΩ	
Inductance	H (henry) mH μH	

CHAPTER 3—FEATURED UNITS 57

Table 3-10.—*USBR preferred units—Electricity and magnetism*—Continued

Quantity	Preferred units—symbol(s) and/or name(s)	Remarks
Magnetic dipole moment	Wb·m (weber meter)	
Magnetic field strength	kA/m A/m	
Magnetic flux	Wb (weber) mWb	
Magnetic induction (flux density)	T (tesla) mT μT	
Magnetic moment	A·m^2	
Magnetic polarization	T mT	
Magnetic potential difference	kA A mA	
Magnetic vector potential	kWb/m Wb/m	
Magnetization	kA/m A/m	
Magnetomotive force	A·t (ampere-turn)	Dimensionally identical to current; the symbol A may be used alone if desired.
Permeability	H/m (henry per meter) mH/m μH/m	
Permeance	H (henry)	
Permittivity	F/m (farad per meter) μF/m nF/m	
Reluctance	H^{-1} (reciprocal henry)	
Resistivity	GΩ·m MΩ·m kΩ·m Ω·m (ohm meter) mΩ·m	

Table 3-11.—*USBR preferred units—Light and electromagnetic radiations*

Quantity	Preferred units—symbol(s) and/or name(s)	Remarks
Illuminance	lx (lux)	
Illumination	klx	
Irradiance	W/m^2 kW/m^2	In meteorology: 1 langley per minute = 697.3 W/m^2
Light exposure	lx·s (lux second)	
Luminance	cd/m^2	
Luminous efficacy	lm/W (lumen per watt)	
Luminous exitance	lm/m^2	
Luminous flux	lm (lumen) mlm	
Luminous intensity	cd (candela) mcd	
Quantity of light	lm·s	
Radiance	W/(sr·m^2)	
Radiant energy	J (joule)	
Radiant exitance	W/m^2	
Radiant flux	W	
Radiant intensity	W/sr	
Radiant power	W (watt)	
Wave length	nm pm	

Table 3-12.—*USBR preferred units—Physical chemistry and molecular physics*

Quantity	Preferred units—symbol(s) and/or name(s)	Remarks
Amount of substance	kmol mol (mole) mmol	
Concentration—		
volume	mg/L mol/L meq/L (milliequivalent per liter) mol/m^3 kmol/m^3	
mass (molality)	mmol/kg mol/kg	
Diffusion coefficient	m^2/s	
Molar entropy	J/(mol·K)	
Molar heat capacity	J/mol·K)	
Molar internal energy	kJ/mol J/mol	
Molar mass	kg/mol g/mol	
Molar volume	m^3/mol L/mol	
Thermal diffusion coefficient	m^2/s	

Table 3-13.—*USBR preferred units—Nuclear physics and ionizing radiations*

Quantity	Preferred units—symbol(s) and/or name(s)	Remarks
Absorbed dose of ionizing radiation	Gy (gray) mGy μGy	
Actvity of radionuclides	TBq GBq MBq kBq Bq (becquerel)	
Angular cross section	m^2/sr	
Atomic attenuation coefficient	m^2	
Cross section	m^2	
Energy flux density	W/m^2	
Half life	s (second)	
Ion number density	m^{-3} (per cubic meter)	
Linear attenuation coefficient	m^{-1} (per meter)	
Mass attenuation coefficient	m^2/kg	
Mass of atom	kg	
Mean free path	m mm	
Mean life	s	
Reaction energy	J (joule)	The electronvolt (eV) is also permitted.

CHAPTER 3—FEATURED UNITS

Table 3-14.—*USBR preferred units—Water resources engineering*

Quantity	Preferred units—symbol(s) and/or name(s)	Remarks
Catchment areas, drainage areas	km² ha m²	Use hectare (ha) and square meter (m²) for small areas.
Channel area	m²	
Channel and pipe lengths	km m	
Depth of application, irrigation	mm	
Discharges, streamflows, and flow volumes	*m³/s m³/d L/s	The AWWA recommends the use of megaliter per day (ML/d) or gigaliter per day (GL/d) in place of mgal/d for municipal and industrial water supply.
Flow gages (laboratory)	L/s L/min m³/s	
Flow velocities	*m/s mm/s	
Gradient (slope)	percent decimal fraction common fraction or ratio	
Hydraulic conductivity—permeability	*m/s m/d (meter per day)	With the appropriate prefix, use meters per second (e.g., nm/s) in soil mechanics and meters per day (e.g., mm/d) for field hydrologic data collection purposes
Hydraulic gradient	percent decimal fraction common fraction	
Hydraulic head	*meters of water (meters-head) kPa	Head is a measure of fluid depth; this can be converted to a pressure value.
Hydraulic pressure	kPa	
Level rod	m	
Pipe area	m² mm²	Small pipes and tubes.
Pipe diameter	mm	

Table 3-14.—*USBR preferred units—Water resources engineering*—Continued

Quantity	Preferred units— symbol(s) and/or name(s)	Remarks
Precipitation—		
rain	mm	Give rainfall intensity in millimeters per hour.
snow	cm	
Reservoir capacity	m^3	Use scientific notation; do not use billion or trillion terminology.
Sample volume	mL	Do not use cubic centimeter.
Sand filter flow	L/(m^2·min)	The AWWA recommends the use of millimeter per second (mm/s) [i.e., mm^3/(mm^2·s)].
Sediment concentration	mg/L	
Sediment load	t/d (metric ton per day) Mg/d (megagram per day)	
Sedimentation basin capacity	L/(m^2·min)	
Specific conductance	μS/cm	For use as a measure of water quality only.
Stream dimensions—		
width	m	
length	km	
cross-sectional area	m^2	
Surface area of body of water	*ha ma	
Water billing, domestic	1000's of liters	Do not use kiloliter (kL) as a substitute. The kiloliter is recommended by the AWWA and is used in Australia.
Water depths	m	
Water levels	m	
Water quality	mg/L	Use as measure of dissolved solids. The use of JTU's, FTU's, and NTU's will remain acceptable units for turbidity.
Water temperature	°C	
Water usage	m^3	

* Indicates unit with greater preference.

CHAPTER 3—FEATURED UNITS 63

Physical and Mental Reference Guides for SI Units

People who use any system of measure do so with an acquired system of mental and physical reference points. The principal reason so many Americans oppose the introduction of the metric system of units is that they will no longer have their acquired reference system; they are not able to relate to or understand the new units. It is estimated that it will require only a minimum of time to reorient one's thinking to the new system; the more someone uses SI measures without converting back and forth, the faster that person will acquire the mental reference system needed to easily use the SI units. The use of SI, which will involve almost everyone in the general population, centers around five familiar quantities: (1) linear measure, (2) area, (3) volume, (4) mass, and (5) temperature.

The following segments and tables are provided to help build a mental reference system for SI units:

The 10-11-12-13 Relationship

For quick approximate conversions relating the meter to the U.S. customary yard, the following relationships may be applied:

$$10 \text{ m} = 11 \text{ yd}$$
$$10 \text{ m}^2 = 12 \text{ yd}^2$$
$$10 \text{ m}^3 = 13 \text{ yd}^3$$

The key to remembering these relationships is the 10-11-12-13 numerical sequence; 10 is the numerical base and the meter is the base unit, the essence of the metric system.

Water Relationship

Water provides a very convenient and reasonably accurate relationship among several SI units of measure. Knowledge of these relationships can be very useful in calculations which do not require a high degree of accuracy. These relationships are summarized in the following statements:

1. One cubic meter (m^3) of water has a mass of 1 metric ton.
2. A liter of water has a mass of 1 kilogram.
3. A liter of water spread over 1 square meter exerts a pressure of 10 pascals. The depth of this 1 liter of water is 1.0 millimeter.
4. A cubic meter of water spread over 1 square meter exerts a pressure of 10 kilopascals.

References for Linear Measurements

(10^{-5} m)   10 μm   —Diameter of red blood cell; twice the thickness of spider web strand

References for Linear Measurements—Continued

(10^{-4} m)	0.1 mm	—Thickness of a sheet of paper
(10^{-3} m)	1.0 mm	—Thickness of a dime; thickness of 10 sheets of paper
(10^{-2} m)	1.0 cm	—Radius of a nickel; joint thickness in brickwork
(10^{-1} m)	0.1 m	—Length of king size cigarette; approximately 4 inches
(10^{0} m)	1.0 m	—Approximately 10 percent longer than a yard; one-half the standard door height
(10^{1} m)	10.0 m	—Twice the height of a giraffe; height of three-story house
(10^{2} m)	100.0 m	—Olympic sprint distance; 10 percent less than a football field which includes the end zones
(10^{3} m)	1.0 km	—Approximately 60 percent of a mile

References for Area Measurements

(10^{-6} m)	1 mm^2	—About the size of the head of a straight pin
(10^{-4} m^2)	1 cm^2	—The center hole of a Lifesaver candy
(10^{-2} m^2)	1 dm^2	—The U.S. dollar bill
(10^{0} m^2)	1 m^2	—Surface area of two card tables; a sheet of AO size drawing paper
(10^{1} m^2)	10 m^2	—Floor area of small bedroom (9 x 12 feet)
(10^{4} m^2)	1 hm^2 (1 ha)	—Field inside Olympic track; field area of two football fields, 2-1/2 acres
(10^{6} m^2)	1 km^2	—Thirty city blocks; slightly larger than three 80-acre farms
(10^{8} m^2)	100 km^2	—Over half the land area of the District of Columbia
(10^{10} m^2)	10 000 km^2	—One-half the land area of New Jersey; the Los Angeles-Long Beach metropolitan area

References for Volume Measurements

(10^{-6} m^3)	1 cm^3 = 1 mL	—A marble
(10^{-3} m^3)	1 L	—Five to six percent larger than a quart
(10^{0} m^3)	1 m^3	—Inside a refrigerator packing box

CHAPTER 3—FEATURED UNITS

References for Volume Measurements—Continued

(10^3 m^3)	1000 m^3	—Volume inside 10 railroad boxcars

References for Mass Measurements

(10^{-4} kg)	100 mg	—Five postage stamps
(10^{-3} kg)	1 g	—Medium raisin
(10^{-2} kg)	10 g	—Two nickels
(10^{-1} kg)	0.1 kg	—D-size flashlight battery
(10^0 kg)	1 kg	—Nine sticks of butter
(10^1 kg)	10 kg	—Large turkey
(10^2 kg)	100 kg	—A professional football linebacker
(10^3 kg)	1 Mg = 1 t	—Almost equal to the long ton; a subcompact automobile

References for Temperatures

Figure 1-3 provides sufficient points of reference when comparing air temperature values given in degrees Celsius versus degrees Fahrenheit.

The following body temperature data are included for your information:

37 °C — Normal temperature
38 °C — Slightly feverish
39 °C — Very feverish
40 °C — Dangerously feverish, equals 104 °F

A simple method for comparing large Celsius temperatures to Fahrenheit temperatures is that the Celsius value is approximately one-half the Fahrenheit value. The error percentage in this approximation is relatively small for Fahrenheit values over 250. For Fahrenheit values less than 250, subtract 30 before dividing by 2; this provides reasonable accuracy until the Fahrenheit values approach minus 40.

Chapter IV

USBR ENGINEERING APPLICATIONS

Introduction

The guidelines and information presented in this chapter are for the purpose of assisting USBR technical personnel in the actual usage of SI units. Information concerning the use of SI units on design drawings and in contract specifications is presented as interim guidelines until final design standards are prepared and approved for use by Bureau personnel.

Many diverse engineering and scientific fields exist within the Bureau of Reclamation, and many conflicting standards presently exist among the specialized technical groups. To promote a smoother transition to SI and ISO practices, and to minimize conflicts between ISO recommendations and presently used standards and technical practices, certain deviations from SI and ISO guidelines will be acceptable for an interim period. These deviations are to be held to a minimum and are not to be extended beyond the time when the associated area of science and engineering adopts SI/ISO standards.

Measurements and Dimensioning in Technical Documents

Specifications

Where practicable, SI metric units should be used in preparing design drawings, specifications, solicitations, and other procurement documents, except for those items of equipment and materials not competitively available in metric measurements.

When the commercial suppliers are capable of competitively furnishing the desired equipment and materials manufactured to metric specifications, the design engineer should specify the metric products in lieu of the products built to U.S. customary dimensions.

When required equipment and materials are not available in SI metric dimensions, or metric items cannot be used successfully as replacement equipment or materials, the equipment and materials shall be specified in U.S. customary units. This also should be done when renovating or replacing old equipment and materials where the metric items are not interchangeable.

This will result in hybrid drawings being used by the Bureau. Though undesirable, this appears to be unavoidable for a period of time.

The use of a dual dimensioning scheme is **generally** to be avoided. However, it may be desirable to employ a dual dimensioning scheme when interfacing between SI design and U.S. customary design. The type of dual dimensioning scheme should be selected by the design engineer.

Generally, SI metric equipment and materials are to be used if they are available, will perform satisfactorily, will allow for competition, and will not result in an abnormally high cost differential. All design engineers should apply the "Rule of Reason" and good engineering judgment. It is not intended that metric materials be used if they will not satisfactorily meet the requirements.

For more detail concerning the use of SI on specification drawings, see the section in this chapter on *Guidelines for SI Design Drawings*, page 125.

Publications

In publications and correspondence other than specifications and procurement documents, it will be standard practice to promote the use of SI units. For metric dimensions and metric measurements originating with SI units, show only the SI units; do not use the U.S. customary equivalents or a dual dimensioning scheme. However, it would not be objectionable to include a proper conversion chart in the appendix.

For dimensions and measurements originating with the U.S. customary units, show the nominal SI equivalent first followed by the U.S. customary units in parentheses.

For drawings used in publications, the use of a dual dimensioning scheme can be more discretionary. The author/editor may elect to superimpose a metric conversion chart showing the nominal metric equivalents of U.S. customary dimensions or use some other scheme; do not show U.S. customary equivalents of SI dimensions. For drawings of materials or hardware, it should be indicated that the nominal SI values represent a soft conversion, and this does not imply that such an item is available with the metric dimensions given.

Contract Specifications

Currently, the Bureau of Reclamation has several construction specifications being prepared in SI units. Listed in table 4-1 are the SI units which will be used for bidding schedule items in metric specifications.

Seven standard specifications paragraphs are presented in the completely metricated format. Presently, not all of these paragraphs are completely metricated; dimensions which are shown as metric but which are currently given in U.S. customary units are underlined; the presently used customary units are listed in tables at the end of each specifications paragraph. The metric values used in the paragraphs are not true soft conversions; most values were taken directly from table 4-2 which lists nominal metric values to be substituted for particular customary values. All standard specifications paragraphs are subject to future review and revision.

CHAPTER 4—USBR ENGINEERING APPLICATIONS

Table 4-1.—*Metric units for bidding schedule items in specifications*

Schedule item	U.S. customary units	SI metric units
Aggregate base and subbase	ton or yd^3	t or m^3
Applying concrete protective coating	yd^2	m^2
Applying liquid asphalt prime coat	ton or gal	t or L
Applying waterbase stabilizer mixture	gal	L
Auger excavation	lin ft	m
Backfill, compacted or pervious	yd^3	m^3
Bedding for riprap or reservoirs	yd^3	m^3
Bonding concrete	yd^2	m^2
Chain link fabric tunnel supports	yd^2	m^2
Clearing	acres	ha
Concrete	yd^3	m^3
Drilling drainage, feeler, or pilot holes	lin ft	m
Drilling grout holes into foundation between the following depths:	lin ft	m
0 to 10 m		
10 to 20 m		
Drilling holes for anchor bars, and grouting bars in place	lin ft	m
Drilling to obtain -mm-diameter[a] cores in concrete	lin ft	m
Earthfill	yd^3	m^3
Epoxy-bonded concrete	yd^2	m^2
Epoxy-bonded epoxy mortar	yd^2	m^2
Excavation	yd^3	m^3
Furnishing and:		
applying soil herbicide	yd^2	m^2
applying sprayed-asphalt emulsion protective coating to foundation surfaces	yd^2	m^2
erecting beam-type guardrail	lin ft	m
erecting chain link fence	lin ft	m
handling cement	cwt	t
handling pozzolan	ton	t

Table 4-1.—*Metric units for bidding schedule items in specifications*—Continued

Schedule item	U.S. customary units	SI metric units
Furnishing and:—Continued		
handling sand bulking materials for pressure grouting	ft^3	m^3
installing electrical conduit or cable	lin ft	m
installing metal seals	lin ft	m
installing miscellaneous metalwork	lb	kg
installing rock bolts	lin ft	m
installing roofing, insulation, tile, or gypsum wallboard	ft^2	m^2
installing structural steel	lb	kg
installing thermocouples	lin ft	m
laying -mm-diameter,[b] No. gage CMP	lin ft	m
laying the following symbol pipe:[b] (list)	lin ft	m
placing elastomeric bearing pads	ft^2	m^2
placing metal grout groove covers	lin ft	m
placing metal pipe and fittings	lb	kg
placing -m-thick riprap	ton or yd^3	t or m^3
placing -mm-outside-diameter[b] metal pipe or tubing, and fittings	lb or ft	kg or m
placing sponge rubber joint filler	ft^2	m^2
placing waterstop	lin ft	m
stringing ACSR conductor or overhead ground wire	mi of line	km of line
Furnishing -mm-diameter[b] perforated pipe and constructing dam embankment drains	lin ft	m
Furnishing soil stabilizer	gal	
Overhaul	mi-yd^3	km·m^3
Placing topsoil	yd^3 or yd^2	m^3 or m^2
Plant-mix bituminous surfacing	ton or yd^3	t or m^3
Pressure grouting	bags	bags[c]

Table 4-1.—*Metric units for bidding schedule items in specifications*—Continued

Schedule item	U.S. customary units	SI metric units
Reinforcing bars	lb	kg
Road maintenance grading	mi	km
Sand, gravel, and cobble drainage blanket	yd^3	m^3
Seeding	acres	ha
Shotcrete protection of excavated surfaces in tunnel	yd^3	m^3
Specially compacted earthfill zone	yd^3	m^3
Steel bearing plates for rock bolts	lb	kg
Stripping	yd^2 or yd^3	m^2 or m^3
Structural steel tunnel supports	lb	kg
Temperature control of concrete	yd^3	m^3
Timber or lumber	Mbm	m^3
Watering	Mgal	L or m^3
Welded wire fabric	ft^2 or lb	m^2 or kg

[a] Core sizes will be given in inches or symbolic measures (e.g., Nx) until metric bits are standardized and available.
[b] Actual pipe diameters (nominal) will be given in inches until nominal metric sizes are standardized. Reference table 4-11.
[c] Retain until manufactured in metric dimensions.

Table 4-2.—*Substituted SI metric values in standard specifications paragraphs* [1]

inches	millimeters	feet	meters
1/16	2	2-1/2	1
1/8	3	4	1.2
3/16	5	6	2
1/4	6	8	2.5
3/8	10	12	3.5
1/2	12		
5/8	15	20	6
3/4	20	25	7.5
7/8	20	30	9
1	25	50	15
1-1/4	30	60	20
1-1/2	40	100	30
1-3/4	45		
2	50	ft^3	m^3
2-1/2	65		
3	75	1	0.03
3-1/2	90		
5	130	lb/in^2 (psi)	kPa
6	150		
8	200	1	7
12	300	50	350
18	450	75	520
20	500	100	700
		200	1400
°F	°C		
		pounds (mass)	kilograms
20	-6		
40	5	5	2
45	8	15	7
55	13	20	10
60	16	94	43
70	21	100	45
75	24		

[1] **CAUTION:** *NOT FOR CONTRACTURAL USE.* Contractors shall not use the above-listed "equivalents" in satisfying contractual requirements. Significant errors would result if the approximate relationships shown were applied to larger dimensions. These "equivalents" represent rationalized nominal dimensions; these values are not exactly equal. Equivalent values for use in construction shall be calculated using the chapter V conversion factors.

CHAPTER 4—USBR ENGINEERING APPLICATIONS

_____. **TOLERANCES FOR CONCRETE CONSTRUCTION**

a. *General.*—Permissible surface irregularities for the various classes of concrete surface finish as specified in Paragraph_____, (Finishes) are defined as "finishes," and are to be distinguished from tolerances as described herein. The intent of this paragraph is to establish tolerances that are consistent with modern construction practice, yet are goverened by the effect that permissible deviations will have upon the structural action or operational function of the structure. Deviations from the established lines, grades, and dimensions will be permitted to the extent set forth herein: <u>Provided</u>, That the Government reserves the right to diminish the tolerances set forth herein if such tolerances impair the structural action or operational function of a structure or portion thereof.

Where specific tolerances are not stated in these specifications or shown on the drawings for a structure, portion of a structure, or other feature of the work, permissible deviations will be interpreted conformably to the tolerances stated in this paragraph for similar work. Specific maximum or minimum tolerances shown on the drawings in connection with any dimension shall be considered as supplemental to the tolerances specified in this paragraph, and shall govern. The contractor shall be responsible for setting and maintaining concrete forms within the tolerance limits necessary to insure that the completed work will be within the tolerances specified. Concrete work that exceeds the tolerance limits specified in these specifications or shown on the drawings shall be remedied or removed and replaced at the expense of and by the contractor.

b. *Tolerances for powerplant structure, pumping plant structure, and buildings.*—

(1) Variation from plumb or specified batter:

    (a) For lines and surfaces of columns, piers, and walls, and for arrises

        In any length of 3000 mm .... 5 mm
        In any length of 6000 mm .... 10 mm
        Maximum for entire length .... 25 mm

Conc. 26M
NOTE TO SPECIFIERS: Subparagraphs b. through i. to be modified, adapted, or deleted as required for the work.

(b) For exposed corner columns, control-joint grooves, and other conspicuous lines

In any length of 6000 mm .... 5 mm
Maximum for entire length .... 15 mm

(2) Variation from level or from grades indicated on the drawings:

    (a) For floors, ceilings, beam soffits, and arrises

In any length of 3000 mm .... 5 mm
In any length of 6000 mm .... 10 mm
Maximum for entire length .... 20 mm

    (b) For exposed lintels, sills, parapets, horizontal grooves, and other conspicuous lines

In any length of 6000 mm .... 5 mm
Maximum for entire length .... 15 mm

(3) Variation of linear building lines from established position in plan and related position of columns, walls, and partitions

In any length of 6000 mm .... 15 mm
Maximum for entire length .... 25 mm

(4) Variation in locations of sleeves, floor openings, and wall openings .... 15 mm

(5) Variation in sizes of floor openings, and wall openings, except wall openings for swinging doors .... 5 mm

(6) Variation in sizes of wall openings for swinging doors

Minus .... 0 mm
Plus .... 5 mm

(7) Variation in cross-sectional dimensions of columns and beams and in the thickness of slabs and walls

Minus .... 5 mm
Plus .... 15 mm

(8) Footings:

    (a) Variation of dimensions in plan

Minus .... 15 mm
Plus .... 50 mm

CHAPTER 4—USBR ENGINEERING APPLICATIONS

(b) Misplacement or eccentricity	2 percent of the footing width in the direction of misplacement but not more than . . . . 50 mm
(c) Reduction in thickness	. . . . 5 percent of specified thickness

(9) Variation in steps:

(a) In a flight of stairs	Rise	3 mm
	Tread	5 mm
(b) In consecutive steps	Rise	2 mm
	Tread	3 mm

c. *Tolerances for substation structures.—*

(1) Variation from plumb or specified batter for lines and surfaces of piers, stems, and walls	In any length of 3000 mm . . . . 5 mm Maximum for entire length . . . . 10 mm
(2) Variation from level or from grades indicated on the drawings for slabs	In any length of 3000 mm . . . . 5 mm In any length of 6000 mm . . . . 10 mm Maximum for entire length . . . . 20 mm
(3) Variation in cross-sectional dimensions of piers and stems and in the thickness of slabs and walls	Minus . . . . 5 mm Plus . . . . 15 mm
(4) Variation from specified elevation for top of concrete for foundations	Minus . . . . 15 mm Plus . . . . 15 mm

Conc. 26M
Sheet 3 of 10

(5) Variation from specified grade or alinement for cable trenches

In any length of 3000 mm .... 5 mm
In any length of 6000 mm .... 10 mm
Maximum for entire length .... 20 mm

(6) Footings:

    (a) Variation of dimensions in plan

    Minus .... 15 mm
    Plus .... 50 mm

    (b) Misplacement or eccentricity

    2 percent of the footing width in the direction of misplacement but not more than .... 50 mm

    (c) Reduction in thickness

    .... 5 percent of specified thickness

d. *Tolerances for canal lining.—*

(1) Departure from established alinement*

.... 50 mm on tangents
.... 100 mm on curves

(2) Departure from established profile grade*

.... 25 mm

(3) Reduction in thickness of lining

.... 10 percent of specified thickness: <u>Provided</u>, That average thickness is not less than specified thickness

(4) Variation from specified width of section at any height

.... 0.25 percent plus 25 mm

CHAPTER 4—USBR ENGINEERING APPLICATIONS

(5) Variation from established height of lining . . . . 0.50 percent plus 25 mm

*Any departure from alinement or grade shall be uniform and no correction in alinement shall be made in less than 30 000 millimeters.

e. *Tolerances for canal structures.—*

  (1) Monolithic siphons and culverts:

    (a) Departure from established alinement . . . . 25 mm

    (b) Departure from established profile grade . . . . 25 mm

    (c) Variation in thickness

      At any point . . . . minus 2.5 percent or 5 mm, whichever is greater

      At any point . . . . plus 5 percent or 15 mm, whichever is greater

    (d) Variation from inside dimensions . . . . 0.50 percent

  (2) Bridges, checks, overchutes, drops, turnouts, inlets, chutes, and similar structures:

    (a) Departure from established alinement . . . . 25 mm

    (b) Departure from established grade . . . . 25 mm

    (c) Variation from plumb or specified batter for lines and surfaces of columns, piers, walls, and for arrises

      Exposed in any length of 3000 mm . . . . 15 mm

      Backfilled, in any length of 3000 mm . . . . 25 mm

(d)	Variation from level or from grades indicated on the drawings for slabs, beams, horizontal grooves, and railing offsets	Exposed in any length of 3000 mm . . . . 15 mm Backfilled, in any length of 3000 mm . . . . 25 mm
(3)	Variation in cross-sectional dimensions of columns, piers, slabs, walls, beams, and similar parts of the structures in (2) above	Minus . . . . 5 mm Plus . . . . 15 mm
(4)	Variation in thickness of bridge slabs	Minus . . . . 3 mm Plus . . . . 5 mm
(5)	Footings:	
	(a) Variation of dimensions in plan	Minus . . . . 15 mm Plus . . . . 50 mm
	(b) Misplacement or eccentricity	2 percent of the footing width in the direction of misplacement but not more than . . . . 50 mm
	(c) Reduction in thickness	. . . . 5 percent of specified thickness
(6)	Variation in sizes and locations of slab and wall openings	. . . . 15 mm
(7)	Variation from plumb or level for sills and sidewalls for radial gates and similar watertight joints**	. . . . not greater than a rate of 0.1 percent
(8)	Variation from plumb of pipe erected vertically	In any length of 3000 mm . . . . 15 mm

**Dimensions between sidewalls for radial gates shall be not more than shown on the drawings at the sills and not less than shown on the drawings at the top of the walls.

CHAPTER 4—USBR ENGINEERING APPLICATIONS

f. *Tolerances for dam structures.—*

(1)	Variation of constructed linear outline from established position in plan	In any length of 6000 mm, except in buried construction . . . . 15 mm Maximum for entire length, except in buried construction . . . . 20 mm In buried construction . . . . . . twice the above amounts
(2)	Variation of dimensions to individual structure features from established positions	Maximum for overall dimension, except in buried construction . . . . 30 mm In buried construction . . . . 65 mm
(3)	Variation from plumb, specified batter, or curved surfaces for all structures, including lines and surfaces of columns, walls, piers, buttresses, arch sections, vertical joint grooves, and visible arrises	In any length of 3000 mm, except in buried construction . . . . 15 mm In any length of 6000 mm, except in buried construction . . . . 20 mm Maximum for entire length, except in buried construction . . . . 30 mm In buried construction . . . . . . twice the above amounts
(4)	Variation from level or from grades indicated on the drawings for slabs, beams, soffits, horizontal joint grooves, and visible arrises	In any length of 3000 mm, except in buried construction . . . . 5 mm Maximum for entire length, except in buried construction . . . . 15 mm In buried construction . . . . . . twice the above amounts
(5)	Variation in cross-sectional dimensions of columns, beams, buttresses, piers, and similar members	Minus . . . . 5 mm Plus . . . . 15 mm

(6) Variation in the thickness of slabs, walls, arch sections, and similar members

Minus .... 5 mm
Plus .... 15 mm

(7) Footings for columns, piers, walls, buttresses, and similar members:

    (a) Variation of dimensions in plan

Minus .... 15 mm
Plus .... 50 mm

    (b) Misplacement or eccentricity

2 percent of the footing width in the direction of misplacement but not more than .... 50 mm

    (c) Reduction in thickness

.... 5 percent of specified thickness

(8) Variation from plumb or level for sills and sidewalls for radial gates and similar watertight joints**

.... not greater than a rate of 0.1 percent

(9) Variation in locations of sleeves, floor openings, and wall openings

.... 15 mm

(10) Variation in sizes of sleeves, floor openings, and wall openings

.... 5 mm

**Dimensions between sidewalls for radial gates shall be not more than shown on the drawings at the sills and not less than shown on the drawings at the top of the walls.

g. *Tolerances for bridges.—*

(1) Departure from established alinement

.... 3 mm

(2) Departure from established grades

.... 3 mm

CHAPTER 4—USBR ENGINEERING APPLICATIONS

(3) Variation from plumb or specified batter for lines and surfaces of piers and walls, and for arrises

Exposed, in any length of 3000 mm . . . . 15 mm
Backfilled, in any length of 3000 mm . . . . 25 mm

(4) Variation from level or from grades indicated on the drawings for slabs, beams, horizontal grooves, railing offsets, and diaphragms

Exposed, in any length of 3000 mm . . . . 15 mm
Backfilled, in any length of 3000 mm . . . . 25 mm

(5) Variation in cross-sectional dimensions of piers, slabs, walls, beams, and similar parts of bridge structures

Minus . . . . 5 mm
Plus . . . . 15 mm

(6) Variation in thickness of bridge slabs

Minus . . . . 3 mm
Plus . . . . 5 mm

(7) Footings:

    (a) Variation of dimensions in plan

Minus . . . . 15 mm
Plus . . . . 50 mm

    (b) Misplacement or eccentricity

2 percent of the footing width in the direction of misplacement but not more than . . . . 50 mm

    (c) Reduction in thickness

. . . . 5 percent of specified thickness

h. *Tolerances for tunnel lining and monolithic conduits.—*

(1) Departure from established alinement or from established grade

Free-flow tunnels and conduits . . . . 25 mm
High-velocity tunnels and conduits . . . . 15 mm

(2) Variation in thickness, at any point

Tunnel lining . . . . minus 0
Conduits . . . . minus 2.5 percent or 5 mm, whichever is greater
Conduits . . . . plus 5 percent or 15 mm, whichever is greater

(3) Variation from inside dimensions

. . . . 0.5 percent

i. *Tolerances for placing reinforcing steel.–*

  (1) Reinforcing steel, except for bridges:

   (a) Variation of protective covering

   With cover of 65 mm or less . . . . 5 mm
   With cover of more than 65 mm . . . . 15 mm

   (b) Variation from indicated spacing

   . . . . 25 mm

  (2) Reinforcing steel for bridges:

   (a) Variation of protective covering

   With cover of 65 mm or less . . . . 3 mm
   With cover of more than 65 mm . . . . 5 mm

   (b) Variation from indicated spacing

   . . . . 25 mm

CHAPTER 4—USBR ENGINEERING APPLICATIONS

_____. FINISHES AND FINISHING

a. *General.*—Allowable deviations from plumb or level and from the alinement, profile grades, and dimensions shown on the drawings are specified in paragraph_____(Tolerances for Concrete Construction), are defined as "tolerances," and are to be distinguished from irregularities in finish as described herein. The classes of finish and the requirements for finishing of concrete surfaces shall be as specified in this paragraph or as indicated on the drawings. The contractor shall notify the contracting officer before finishing concrete. Unless inspection is waived in each specific case, finishing of concrete shall be performed only in the presence of a Government inspector. Concrete surfaces will be tested by the Government where necessary to determine whether surface irregularities are within the limits hereinafter specified.

Surface irregularities are classified as "abrupt" or "gradual." Offsets caused by displaced or misplaced form sheathing or lining or form sections, or by loose knots in forms or otherwise defective form lumber will be considered as abrupt irregularities, and will be tested by direct measurements. All other irregularities will be considered as gradual irregularities, and will be tested by use of a template, consisting of a straightedge or the equivalent thereof for curved surfaces. The length of the template will be 1.5 meters for testing of formed surfaces and 3 meters for testing of unformed surfaces.

b. *Formed surfaces.*—The classes of finish for formed concrete surfaces are designated by use of symbols [1] (F1, F2, F3, and F4). No sack rubbing or sandblasting will be required as a finish on formed surfaces. No grinding will be required on formed surfaces, other than that necessary for repair of surface imperfections. Unless otherwise specified or indicated on the drawings, the classes of finish shall apply as follows:

(1) F1.—Finish F1 applies to formed surfaces upon or against which fill material or concrete is to be placed. The surfaces require no treatment after form removal except for repair of defective concrete and filling of holes left by the removal of fasteners from the

[1] Delete or revise as required.

ends of tie rods as required in paragraph _____ (Repair of Concrete) and the specified curing. Correction of surface irregularities will be required for depressions only, and only for those which, when measured as described in subparagraph a., exceed 25 millimeters.

(2) F2.—Finish F2 applies to all formed surfaces not permanently concealed by fill material or concrete, or not required to receive finish [1] (F3 or F4).

[2] [Surfaces to receive finish F2 include:

(a) _____
_____.

(b) _____
_____.

(c) _____
_____.]

Surface irregularities, measured as described in subparagraph a., shall not exceed 5 millimeters for abrupt irregularities and 15 millimeters for gradual irregularities [1] (: Provided, That surfaces over which radial gate seals will operate without sill or wallplates shall be free from abrupt irregularities.)

(3) F3.—Finish F3 applies to formed surfaces, the appearance of which is considered by the Government to be of special importance, such as surfaces of structures prominently exposed to public inspection.

[2] Delete or insert specific surfaces. Surfaces of canal and lateral structures, including inside surfaces of siphons, culverts, and tunnel linings; structures appurtenant to earth dams, including surfaces of outlet works and open spillways; small pumping plants and powerplants; bridges and retaining walls not prominently exposed to public inspection; galleries and tunnels in dams; and concrete dams are normally required to receive finish F2.

CHAPTER 4—USBR ENGINEERING APPLICATIONS

³ [Surfaces to receive finish F3 include:

(a) _____
    _____.

(b) _____
    _____.

(c) _____
    _____.]

Surface irregularities, measured as described in subparagraph a., shall not exceed 5 millimeters for gradual irregularities and 3 millimeters for abrupt irregularities, except that abrupt irregularities will not be permitted at construction joints.

(4) F4.—Finish F4 applies to formed surfaces for which accurate alinement and evenness of surface are of paramount importance from the standpoint of preventing destructive effects of water action.

⁴ [Surfaces to receive finish F4 include:

(a) _____
    _____.

(b) _____
    _____.

(c) _____
    _____.]

³Insert specific surfaces. Surfaces of superstructures of large pumping plants and powerplants; parapets, railings, and decorative features on dams and bridges; and permanent buildings are normally required to receive finish F3.
⁴Insert specific surfaces. Surfaces of suction tubes, draft tubes, and the lower semicircumference of tunnel spillways are normally required to receive finish F4.

When measured as described in subparagraph a., abrupt irregularities shall not exceed 5 millimeters for irregularities parallel to the direction of flow, and 3 millimeters for irregularities not parallel to the direction of flow. Gradual irregularities shall not exceed 5 millimeters.

c. *Unformed surfaces.*—The classes of finish for unformed concrete surfaces are designated by the symbols [1] (U1, U2, U3, and U4). Interior surfaces shall be sloped for drainage where shown on the drawings or directed. Surfaces which will be exposed to the weather and which would normally be level, shall be sloped for drainage. Unless the use of other slopes or level surfaces is indicated on the drawings or directed, narrow surfaces, such as tops of walls and curbs, shall be sloped approximately 1:30 of width; and broader surfaces, such as walks, roadways, platforms, and decks shall be sloped approximately 1:50. Unless otherwise specified or indicated on the drawings, these classes of finish shall apply as follows:

(1) U1.—Finish U1 (screeded finish) applies to unformed surfaces that will be covered by fill material or by concrete, [1] (surfaces of operating platforms on canal structures, and surfaces of subfloors which will be covered by concrete floor topping.) Finish U1 is also used as the first stage of finishes U2 and U3. Finishing operations shall consist of sufficient leveling and screeding to produce even uniform surfaces. Surface irregularities, measured as described in subparagraph a., shall not exceed 10 millimeters.

(2) U2.—Finish U2 (floated finish) applies to unformed surfaces not permanently concealed by fill material or concrete, or not required to receive finish [1] (U1, U3, or U4).

[5] [Surfaces to receive finish U2 include:

[5] Delete or insert specific surfaces. Surfaces of canal structures, including the inverts of siphons and tunnels; floors of spillways, outlet works, and stilling basins; outside decks of pumping plants and powerplants; slabs to be covered with built-up roofing or membrane waterproofing; floors of service tunnels, sumps, culverts, and temporary diversion conduits; tops of transmission line and bridge piers, and tops of walls; and surfaces of gutters, sidewalks, and outside entrance slabs are normally required to receive finish U2.

CHAPTER 4—USBR ENGINEERING APPLICATIONS

(a)_____
_____ .

(b)_____
_____ .

(c)_____
_____ .]

Finish U2 is also used as the second stage of finish U3. Floating may be performed by use of hand- or power-driven equipment. Floating shall be started as soon as the screeded surface has stiffened sufficiently, and shall be the minimum necessary to produce a surface that is free from screed marks and is uniform in texture. If finish U3 is to be applied, floating shall be continued until a small amount of mortar without excess water is brought to the surface, so as to permit effective troweling. Surface irregularities, measured as described in subparagraph a., shall not exceed 5 millimeters, [1] (except that surface irregularities on roadways of bridge decks shall not exceed 3 millimeters). Joints and edges of gutters, sidewalks, and entrance slabs, and other joints and edges shall be tooled where shown on the drawings or directed.

[6] (At the proper interval after being struck off, the roadway slabs of concrete bridges shall be finished by wood floating or belting to produce a nonskid surface equivalent to that obtainable by use of the best modern practice in finishing pavements for highways.)

[7] (After the surfaces of roadway slabs of concrete bridges have been wood floated, the surfaces shall be given a broom finish. The finish shall be applied when the water sheen has practically disappeared. The broom shall be drawn transversely across the pavement with adjacent strokes slightly overlapping. The brooming shall be completed before the concrete is in such condition that the surface will be torn or unduly roughened by the operation. The

---

[6] Include only for roadway slabs of bridge decks that will be covered with a sealant.
[7] Include only for roadway slabs of bridge decks that will not be covered with a sealant.

finished surface shall have a uniform appearance and shall be free of corrugations exceeding 3 millimeters in depth. Brooms shall be of a quality, size, and construction, and be so operated, as to produce a surface finish satisfactory to the contracting officer.)

(3) U3.—Finish U3 (troweled finish) applies to ¹(inside floors of buildings, except floors requiring a bonded-concrete finish or a terrazzo finish, and to inverts of draft tubes and tunnel spillways.) When the floated surface has hardened sufficiently to prevent an excess of fine material from being drawn to the surface, steel troweling shall be started. Steel troweling shall be performed with firm pressure so as to flatten the sandy texture of the floated surface and produce a dense uniform surface, free from blemishes and trowel marks. Surface irregularities, measured as described in subparagraph a., shall not exceed 5 millimeters.

(4) U4.—Finish U4 applies to canal and lateral linings. The finished surface shall be equivalent in evenness, smoothness, and freedom from rock pockets and surface voids to that obtainable by effective use of a long-handled steel trowel. Light surface pitting and light trowel marks will not be considered objectionable. Where the surface produced by a lining machine meets the specified requirements, no further finishing operations will be required. Surface irregularities, measured as described in subparagraph a., shall not exceed 5 millimeters for bottom slabs and 15 millimeters for side slopes.

10-15-73 Revisions: Added requirements for finishing roadway slabs of bridge decks that will not be covered with a sealant. Other minor revisions.

CHAPTER 4—USBR ENGINEERING APPLICATIONS

_____. COARSE AGGREGATE

a. *General.*—The term "coarse aggregate," for the purpose of these specifications, designates aggregate of sizes within the range of 4.75 millimeters to [1] (38.1 millimeters) or any size or range of sizes within such limits. The coarse aggregate shall be reasonably well graded within the nominal size ranges hereinafter specified. Coarse aggregate for concrete shall be furnished by the contractor [2] (from any approved source as provided in subparagraph d.) and shall consist of natural gravel or crushed rock or a mixture of natural gravel and crushed rock. [2] (If crushed coarse aggregate is used with natural coarse aggregate, the crushed aggregate shall be blended uniformly with the uncrushed aggregate. Crushing and blending operations shall at all times be subject to approval.)

[2] (Any royalties or other charges required to be paid for materials taken from deposits not owned by the Government and controlled by the Bureau of Reclamation shall be paid by the contractor.)

Coarse aggregate, as delivered to the batching plant, shall have a uniform and stable moisture content.

b. *Quality.*—The coarse aggregate shall consist of clean, hard, dense, durable, uncoated rock fragments. The percentages of deleterious substances in any size of coarse aggregate, as delivered to the mixer, shall not exceed the following values:

NOTE TO SPECIFIERS: For concrete work in the Kansas-Nebraska area, special changes and additions to this paragraph will be necessary. When preparing specifications for concrete work in this area, contact Code 1511, Engineering and Research Center, for specific requirements.

[1] When the maximum-size aggregate is to be other than 38.1 millimeters, substitute maximum size to be used.
[2] Delete when "Production of Sand and Coarse Aggregate" paragraph is used.

	Percent by mass
Material passing 75-μm (No. 200) screen (designation 16)	0.5
Lightweight material (designation 18)	2
Clay lumps (designation 13)	0.5
Other deleterious substances	1

The sum of the percentages of all deleterious substances in any size, as delivered to the mixer, shall not exceed 3 percent, by mass. Coarse aggregate may be rejected if it fails to meet the following test requirements:

(1) Los Angeles rattler test (designation 21).—If the loss, using grading A, exceeds 10 percent, by mass, at 100 revolutions or 40 percent, by mass, at 500 revolutions.

(2) Sodium-sulfate test for soundness (designation 19).—If the weighted average loss after 5 cycles is more than 10 percent, by mass.

(3) Mass density (designation 10).—If the mass density (saturated surface-dry basis) is less than 2600 kilograms per cubic meter.

The designations in parentheses refer to methods of test described in the Eighth Edition of the Bureau of Reclamation Concrete Manual.

c. *Separation.*—The coarse aggregate shall be separated into nominal sizes and shall be graded, as batched, as follows:

CHAPTER 4—USBR ENGINEERING APPLICATIONS

Designation of size	Nominal size range	Minimum percent retained on screens indicated
19.0 mm	4.75-19.0 mm	50 percent on 9.5 mm
38.1 mm	19.0-38.1 mm	25 percent on 31.5 mm
[3] (75 mm)	38.1-75 mm	25 percent on 63.5 mm)

[4] (Coarse aggregate shall be finished screened at the batching plant on vibrating screens: Provided, That the contractor may use stationary sloping screens having slotted openings if the crushed and broken portion of the total coarse aggregate, as determined by the contracting officer, is less than 30 percent, by mass. Minus 4.75-millimeter material shall be wasted as directed, or may be used to make up deficiencies in the natural sand grading: Provided, That the material, if used, shall be routed back through the classifier in uniform increments for uniform blending with the natural sand being processed: Provided further, That the addition of the minus 4.75-millimeter material shall not cause variations in or exceed the specified grading of the sand as batched. The method and rate of feed shall be such that the screens will not be overloaded and will operate properly in a manner that will result in a finished product which consistently meets the grading requirements of these specifications. The vibrating screens shall be mounted on the batching plant, or at the option of the contractor, the screens may be mounted on the ground adjacent to the batching plant. The vibrating screens shall be so mounted that the vibration of the screens will not be transmitted to or affect the accuracy of the batching scales. The finished products shall pass directly to the individual batching plant storage bins. Stationary slotted sloping screens, where permitted, shall meet the following requirements:

[3] Delete requirements if 75-mm-maximum-size aggregate is not used.

[4] Delete one or both of provisions [4] and [6], as applicable. Both provisions will be deleted when the amount of concrete involved under the specifications being prepared is 400 cubic meters or less. The provision for vibrating screens with the optional use of slotted stationary sloping screens under specific conditions will generally be used when the amount of concrete involved under the specifications being prepared is more than 400 and less than 4000 cubic meters. The provision for finish screening on vibrating screens without the option for slotted stationary sloping screens will generally be used when the amount of concrete involved under the specifications being prepared is 4000 cubic meters or more.

Conc. 7M

(1) The screens shall be at least 2 meters in length and shall be not less than 1.2 meters wide. The flow of the material shall be perpendicular to the narrow dimension of the slots.

(2) The slots shall be not more than 100 millimeters long.

(3) The slope of the screens shall be not less than 30° and not more than 50° from the horizontal, and shall be readily adjustable within that range.)

[5] (The above requirements for finish screening at the batching plant will be waived by the contracting officer if the contractor furnishes bins or other suitable facilities for handling and storing the materials, after they have been separated into nominal sizes, to assure that no objectionable breakage, segregation, or contamination of the materials has taken place, and if the materials as batched fully meet the requirements of these specifications for coarse aggregate.)

[6] (Coarse aggregate shall be finished screened on vibrating screens mounted on the batching plant, or at the option of the contractor, the screens may be mounted on the ground adjacent to the batching plant. The finish screens, if installed over the batching plant, shall be so mounted that the vibration of the screens will not be transmitted to, or affect the accuracy of the batching scales. The method and rate of feed shall be such that the screens will not be overloaded and will operate properly in a manner that will result in a finished product which consistently meets the grading requirements of these specifications. The finished products shall pass directly to the individual batching bins. Minus 4.75-millimeter material shall be wasted as directed or may be used to make up deficiencies in the natural sand grading: Provided, That the material, if used, shall be routed back through the classifier in uniform increments for uniform blending with the natural sand being processed: Provided further, That the addition of the minus 4.75-millimeter material shall not cause variations in or exceed the specified grading of the sand as batched.)

---

[5] Use this waiver only in specifications for transmission lines.
[6] See Note [4]

CHAPTER 4—USBR ENGINEERING APPLICATIONS

Separation of the coarse aggregate into the specified sizes, after finish screening, shall be such that, when the aggregate, as batched, is tested by screening on the screens designated in the following tabulation, the material passing the undersize test screen (significant undersize) shall not exceed [7] (3 percent, 2 percent,) by mass, and all material shall pass the oversize test screen:

Size	Size of square opening in screen	
	For undersize test	For oversize test
19.0-mm aggregate	4.0 mm	26.4 mm
38.1-mm aggregate	16.0 mm	45 mm
[3] (75-mm aggregate	31.5 mm	90 mm)

Screens used in making the tests for undersize and oversize will conform to ASTM Designation: E 11, with respect to permissible variations in average openings.

[2] [d. *Test and approval.*—If coarse aggregate is to be obtained from a deposit not previously tested and approved by the Government, the contractor shall submit representative samples for preconstruction test and approval at least 45 days before the coarse aggregate is required for use. The samples shall consist of approximately 100 kilograms of the 4.75- to 19.0-millimeter coarse aggregate and 50 kilograms of [8] (the 19.0- to 38.1-millimeter coarse aggregate.)

The approval of deposits by the contracting officer shall not be construed as constituting the approval of all or any specific materials taken from the deposits, and the contractor will be held responsible for the specified quality of all such materials used in the work.

Conc. 7M
Sheet 5 of 6

[7] Delete one of these percentages as applicable. Specify "3 percent" when screening over vibrating screens, or slotted stationary sloping screens, as permitted, is specified and "2 percent" when finish screening over vibrating screens, without the option for slotted stationary sloping screens, is specified.
[8] When the maximum-size coarse aggregate is to be larger than 38.1 millimeters, change this portion of sentence to read "each of the other sizes of coarse aggregate."

In addition to preconstruction test and approval of the deposit, the Government will test the coarse aggregate during the progress of the work and the contractor shall provide such facilities as may be necessary for procuring representative samples.]

7-1-74 Revisions: Revised wording in provisions [4] and [6] on sheets 3 and 4.

CHAPTER 4—USBR ENGINEERING APPLICATIONS

_____. SAND

a. *General.*—The term "sand" is used to designate aggregate in which the maximum size of particles is 4.75 millimeters. Sand for concrete, mortar, and grout shall be furnished by the contractor [1] (from any approved source as provided in subparagraph d.) and shall be natural sand, except that crushed sand may be used to make up deficiencies in the natural sand grading. [1] (Crushed sand, if used to make up deficiencies in the natural sand grading shall be produced by a suitable ball or rod mill, disk, or cone crusher, or other approved equipment, so that the sand particles shall be predominantly cubical in shape and free from objectionable quantities of flat or elongated particles. Crusher fines produced by a jaw crusher that will pass a screen having 4.75-millimeter square openings shall be wasted. The crushed sand shall be blended uniformly with the natural sand. Crushing and blending operations shall at all times be subject to approval.)

[1] (Any royalties or other charges required to be paid for materials taken from deposits not owned by the Government and controlled by the Bureau of Reclamation, shall be paid by the contractor.)

Sand, as delivered to the batching plant, shall have a uniform and stable moisture content.

b. *Quality.*—The sand shall consist of clean, hard, dense, durable, uncoated rock fragments. The maximum percentages of deleterious substances in the sand, as delivered to the mixer, shall not exceed the following values:

NOTE TO SPECIFIERS: For concrete work in the Kansas-Nebraska area, special changes and additions to this paragraph will be necessary. When preparing specifications for concrete work in this area, contact Code 1511, Engineering and Research Center, for specific requirements.

[1] Delete when "Production of sand and coarse aggregate" paragraph is used.

	Percent by mass
Material passing 75-μm (No. 200) screen (designation 16)	3
Lightweight material (designation 17)	2
Clay lumps (designation 13)	1
Total of other deleterious substances (such as alkali, mica, coated grains, soft flaky particles, and loam)	2

The sum of the percentages of all deleterious substances shall not exceed 5 percent, by mass. Sand producing a color darker than the standard in the colorimetric test for organic impurities (designation 14) may be rejected. Sand having a mass density (designation 9, saturated surface-dry basis) of less than 2600 kilograms per cubic meter may be rejected. The sand may be rejected if the portion retained on a 300-micrometer (No. 50) screen, when subjected to 5 cycles of the sodium-sulfate test for soundness (designation 19), shows a weighted average loss of more than 8 percent, by mass. The designations in parentheses refer to methods of tests described in the Eighth Edition of the Bureau of Reclamation Concrete Manual.

c. *Grading.*—The sand as batched shall be well graded, and when tested by means of standard screens (designation 4), shall conform to the following limits:

CHAPTER 4—USBR ENGINEERING APPLICATIONS

Screen size	Individual percent by mass retained on screen
4.75 mm (No. 4)	0 to 5
2.36 mm (No. 8)	5 to 15*
1.18 mm (No. 16)	10 to 25*
600 μm (No. 30)	10 to 30
300 μm (No. 50)	15 to 35
150 μm (No. 100)	12 to 20
Pan	3 to 7

*If the individual percent retained on the 1.18-mm (No. 16) screen is 20 percent or less, the maximum limit for the individual percent retained on the 2.36-mm (No. 8) screen may be increased to 20 percent.

[1](d. *Test and approval.*—If sand is to be obtained from a deposit not previously tested and approved by the Government, the contractor shall submit representative samples for preconstruction test and approval at least 45 days before the sand is required for use. The samples shall consist of approximately 100 kilograms of sand.

The approval of deposits by the contracting officer shall not be construed as constituting the approval of all or any specific materials taken from the deposits, and the contractor will be held responsible for the specified quality of all such materials used in the work.

In addition to preconstruction test and approval of the deposit, the Government will test the sand during the progress of the work and the contractor shall provide such facilities as may be necessary for procuring representative samples.)

Conc. 6M
Sheet 3 of 3

8-15-73 Revisions: Added requirements for crushed sand in subparagraph a. Other minor revisions.

____. POLYVINYL-CHLORIDE WATERSTOPS

a. *General.*—PVC (polyvinyl-chloride) waterstops shall be placed in [1] (construction, contraction, and expansion joints of the _____

_____

_____)

where shown on the drawings or where directed.

The contractor shall furnish the waterstops and all materials and equipment for splicing waterstops, for fastening waterstops to the forms and to the supporting reinforcing bars, and for completing the installation of the waterstops. The contractor shall provide suitable support and protection for the waterstops during the progress of the work and shall repair or replace at his expense any damaged waterstops which, in the opinion of the contracting officer, have been damaged to such an extent as to affect the serviceability of the waterstops. All waterstops shall be protected from oil, grease, and curing compound.

[1] [b. *Drawings and data to be furnished by the contractor.*—At least 60 days prior to installing any waterstop, the contractor shall submit drawings and data to the Government for approval.

(1) *Drawings.*—The contractor shall submit four sets of drawings showing details of the waterstops, including dimensions, shapes, and details of intersections and splices between waterstops of the same sizes and of different sizes. The details of intersections and splices shall show all expected connections required for the work under these specifications.

One set of drawings will be returned to the contractor either approved, not approved, or conditionally approved, and also marked to indicate changes, if-required. All drawings that are not approved or that require changes shall be revised and resubmitted for approval, and shall show all changes with revision dates. Drawings conditionally approved shall be resubmitted for approval, if directed.

[1] Delete or revise as required. Designers will determine if drawings are to be required for approval. Data should normally be required unless quantities of waterstops are small.

C-902M

CHAPTER 4—USBR ENGINEERING APPLICATIONS

Any fabrication or procurement of materials performed prior to approval of the drawings will be at the contractor's risk. The Government shall have the right to require the contractor to make any changes in his drawings which may be necessary to make the finished installation conform to the requirements and intent of these specifications without additional cost to the Government. Approval by the Government of the contractor's drawings shall not be held to relieve the contractor of any part of the contractor's obligations to meet all of the requirements of these specifications or of the responsibility for the correctness of the contractor's drawings.

(2) Data.—The contractor shall submit detailed laboratory test reports on the physical properties, listed in subparagraph c., of the compound which will be used in the waterstops to be furnished, together with a copy of the purchase order for the waterstops, and a manufacturer's certificate stating that the waterstops as furnished will meet all requirements of these specifications. If the contractor purchases the waterstops under more than one purchase order, the data and samples required by this subparagraph shall be submitted for each separate purchase.

(3) Addresses for submittals.—The approval drawings and data shall be forwarded to the Bureau of Reclamation, Engineering and Research Center, P.O. Box 25007, Denver Federal Center, Denver, Colorado 80225, Attention: Codes _____ and _____. A copy of each letter transmitting the approval drawings and data shall be forwarded by the contractor to the Construction Engineer, Bureau of Reclamation, _____

_____
_____.

(4) Cost.—The entire cost of preparing and submitting the drawings and data and of furnishing the required samples of finished waterstops shall be included in the applicable unit price per metre bid in the schedule for furnishing and placing waterstops.]

c. *Material.*—PVC waterstops shall be fabricated from a compound, the basic resin of which shall be domestic virgin PVC. No reclaimed PVC or manufacturer's scrap shall be used. The compound shall contain any additional resins, plasticizers, stabilizers, or other materials needed to insure that, when the material is compounded, the finished product will have the following physical characteristics:

Type of test	Method of test	Required
Tensile strength, minimum	ASTM Designation: D 638, speed D, specimen type IV	13.8 MPa
Ultimate elongation, percent minimum	ASTM Designation: D 638, speed D, specimen type IV	300
Stiffness in flexure, minimum	ASTM Designation: D 747	4.1 MPa
Low temperature brittleness at minus 37 °C	ASTM Designation: D 746	No cracking or chipping
Volatile loss, change in mass, percent allowed	ASTM Designation: D 1203, method A, 2-millimeter thick specimen	0.50
Tensile strength after accelerated extraction test, percent of tensile strength before extraction test, minimum	U.S. Army Corps of Engineers Specification CRD-C-572-74	80

C-902M

CHAPTER 4—USBR ENGINEERING APPLICATIONS

Type of test	Method of test	Required
Ultimate elongation after accelerated extraction test, percent of ultimate elongation before extraction test, minimum	U.S. Army Corps of Engineers Specification CRD-C-572-74	80
Change in mass after effect of alkalies test, percent allowed	U.S. Army Corps of Engineers Specification CRD-C-572-74	+0.25, −0.10
Change in Shore Durometer hardness after effect of alkalies test, percent allowed	U.S. Army Corps of Engineers Specification CRD-C-572-74	±5

d. *Fabrication.*—All waterstops shall be molded or extruded in such a manner that any cross section will be dense, homogeneous, and free from porosity and other imperfections.

The cross section of the waterstop shall be uniform along its length and the thickness shall be symmetrical transversely. Tolerances from the dimensions [1] (specified below) (shown on the drawing) shall be plus or minus 5 millimeters in width, plus or minus 2 millimeters in thickness, as [1] (specified below) (shown on the drawing) for rib height, and plus or minus 1 millimeter for other dimensions.

[2] [The type _____ , _____ , and _____ waterstops shall be fabricated in accordance with detail

C-902M
Sheet 4 of 7

[2] Include when specifications will include drawing showing the dimensions and tolerances for the waterstops.

dimensions and tolerances shown on drawing No. _____ (_____).]

[3] (The 225-millimeter, type A, waterstop shall have a center bulb of 20-millimeter inside diameter and 30-millimeter outside diameter; shall be 225 millimeters in width; shall have a minimum of eight longitudinal ribs on each side of the bulb, with the ribs evenly distributed between the rib adjacent to the bulb and the edge of the waterstop, and with 3-millimeter minimum to 6-millimeter maximum rib height; and shall have a web thickness of 10 millimeters adjacent to the center bulb and a web thickness of 6 millimeters near the edge.)

The 150-millimeter, type B, waterstop shall have a center bulb of 6-millimeter inside diameter and 12-millimeter outside diameter; shall be 150 millimeters in width; shall have a minimum of six longitudinal ribs on each side of the bulb, with the ribs evenly distributed between the rib adjacent to the bulb and the edge of the waterstop, and with 3-millimeter minimum to 6-millimeter maximum rib height; and shall have a web thickness of 10 millimeters adjacent to the center bulb and a web thickness of 6 millimeters near the edge.

The _____ -millimeter waterstop shall have a center bulb of _____ -millimeter inside diameter and _____ -millimeter outside diameter; shall be _____ millimeters in width; shall have a minimum of _____ longitudinal ribs on each side of the bulb, with the ribs evenly distributed between the rib adjacent to the bulb and the edge of the waterstop, and with _____ -millimeter minimum to _____ -millimeter maximum rib height; and shall have a web thickness of _____ millimeters adjacent to the center bulb and a web thickness of _____ millimeters near the edge.)

e. *Inspection and tests.—*

(1) General.—All waterstops shall be sampled at the jobsite, tested, and approved by the Government before installation. The contractor shall have the waterstop available at the jobsite in sufficient time to allow 30 days for testing after samples obtained by the Government have been received in the Denver Laboratories.

[3] Include, as required, when specifications will not include drawings showing the details and dimensions of the waterstops.

CHAPTER 4—USBR ENGINEERING APPLICATIONS

(2) *Samples for tests.*—A representative sample not less than 300 millimeters long shall be cut from each 150 meters of each size and type of finished waterstop: Provided, That a minimum of four samples shall be taken for each size and type from each separate purchase order. Each sample shall be marked so that it may be identified with the specific length of waterstop from which it is taken.

(3) *Methods of tests.*—Test specimens will be prepared from the samples in accordance with the U.S. Army Corps of Engineers Specification CRD-C-572-74, single copies of which may be obtained free of charge from the Director, U.S. Army Engineer Waterways Experiment Station, P.O. Box 631, Vicksburg, Mississippi 39180. Tests will be made in accordance with the methods specified in subparagraph c.

f. *Installation.*—Waterstops shall not be installed until [1] (drawings, data, and) field-sampled materials have been approved. The location and embedment of waterstops shall be as shown on the drawings, with approximately one-half of the width of the waterstop embedded in the concrete on each side of the joint. In order to eliminate faulty installation that may result in joint leakage, particular care shall be taken that the waterstops are correctly positioned and secured during installation. All waterstops shall be installed so as to form a continuous watertight diaphragm in the joint unless otherwise shown.

Adequate provision shall be made to completely protect the waterstops during the progress of the work.

Concrete surrounding the waterstops shall be given additional vibration, over and above that used for adjacent concrete placement, to assure complete embedment of the waterstops in the concrete. Larger pieces of aggregate near the waterstops shall be removed by hand during embedment to assure complete contact between the waterstop and the surrounding concrete.

Prior to starting installation of the waterstops, the contractor shall furnish to the Construction Engineer, Bureau of Reclamation, _____
_____
_____,
a copy of the manufacturer's recommendations for installing and making splices in the waterstops. Splices of waterstops shall be fabricated only by workmen who have demonstrated to the satisfaction of the contracting officer that they are sufficiently skilled to fabricate the required splices. Splices in the continuity or at the intersections of runs of plastic waterstops shall be performed by heat sealing the adjacent surfaces in accordance with the manufacturer's recommendations. A thermostatically controlled electric heat source shall be used to make all splices. The correct temperature at which splices should be made will differ with the materials compounded but should be sufficient to melt but not char the plastic material. All splices shall be neat with the ends of the joined waterstops in true alinement. A miter-box guide and portable saw shall be provided and used to cut the ends to be joined to insure good alinement and contact between joined surfaces. The spliced area, when cooled and bent by hand to as sharp an angle as possible, shall show no sign of separation.

[1] [Where splices are required between waterstops of different sizes, the splices shall be made as recommended by the manufacturer of the waterstops and drawings showing the details of the splices shall be submitted to the Government, for approval, as required in subparagraph b.(1) above.]

g. *Measurement and payment.*—Measurement, for payment, of the various types of plastic waterstops will be made of the number of meters of waterstops in place measured along the centerline of the waterstop, with no allowance for lap at splices and intersections.

Payment for furnishing and placing the various types of waterstops will be made at the applicable unit price per meter bid therefor in the schedule, which price shall include the cost of furnishing all material, making field splices and intersections, and installing the waterstops.

Table 4-3.—*Actual customary values for underlined SI metric values for polyvinyl-chloride waterstops specification paragraph (C-902M). These are listed in order of appearance.*

Proposed metric	Actual customary
13.8 MPa	2000 lb/in^2
4.1 MPa	600 lb/in^2
mass	weight
2-mm	0.08-inch
mass	weight
5 mm	3/16 inch
2 mm	1/16 inch
1 mm	1/32 inch
225-mm	9-inch
20-mm	3/4-inch
30-mm	1-1/4-inch
225 mm	9 inches
3-mm	1/8-inch
6-mm	1/4-inch
10 mm	3/8 inch
6 mm	1/4 inch
150-mm	6-inch
6-mm	1/4-inch
12-mm	1/2-inch
150 mm	6 inches
3 mm	1/8 inch
6 mm	1/4 inch
10 mm	3/8 inch
6 mm	1/4 inch
___ millimeter(s) (8 places)	___ inch(es)

___. **RUBBER WATERSTOPS**

a. *General.*—Rubber waterstops shall be placed in [1] [construction, contraction, and expansion joints of the_____

_____

where shown on the drawings or where directed: Provided, That PVC waterstops conforming to paragraph _____ (Polyvinyl-Chloride Waterstops) may be furnished and placed in lieu of rubber waterstops.]

[1] [Type "A," "B," "G," and "H" rubber waterstops shall be in accordance with the details shown on drawing No. _____ (40-D-2867). Type "D," "E," and "F" rubber waterstops shall be in accordance with the details shown on drawing No. _____ (40-D-4567).]

The contractor shall furnish the waterstops and all materials and equipment for splicing waterstops, for fastening waterstops to the forms and to the supporting reinforcing bars, and for completing the installation of the waterstops. The contractor shall provide suitable support and protection for the waterstops during the progress of the work and shall repair or replace at his expense any damaged waterstops which, in the opinion of the contracting officer, have been damaged to such an extent as to affect the serviceability of the waterstops. All waterstops shall be protected from oil, grease, and curing compound.

b. *Material.*—

(1) Rubber waterstop.—The rubber waterstops shall be fabricated from a high-grade, tread-type compound. The basic polymer shall be natural rubber or a synthetic rubber. The material shall be compounded and cured to have the following physical characteristics:

C-888M

[1] Delete or revise as required.

CHAPTER 4—USBR ENGINEERING APPLICATIONS

Type of test	Method of test ASTM Designation	Required Natural rubber	Required Synthetic rubber
Tensile strength, minimum	D 412	24.1 MPa	*20.7 MPa
Tensile strength at 300 percent modulus, minimum	D 412	10 MPa	7.9 MPa
Elongation at break, percent, minimum	D 412	500	*450
Shore durometer (type A)	D 2240	60 to 70	60 to 70
Change in mass, water immersion, percent maximum (2 days at 70 °C)	D 471	5	5
Compression set (constant deflection) percent of original deflection, maximum	D 395, Method B	30	30
Accelerated aging (96 hours at 70 °C) percent of tensile strength before aging, minimum	D 573	80	80
Percent of elongation before aging, minimum	D 573	80	80
Ozone cracking resistance (7 days at 0.5 mg/ℓ at 38 °C) 20 percent elongation	D 1149	No cracks	No cracks

*Polychloroprene shall have a minimum tensile strength of 13.8 megapascals and minimum elongation of 350 percent.

(2) *Gum rubber and rubber cement.*—Gum rubber and rubber cement shall be suitable for making field connections in rubber waterstops as described in succeeding subparagraph _____ (Installation).

[1] (3) *Connection plates.*—Connection plates shall be made from No. 16 United States Standard gage stainless steel plates. The stainless steel plates shall be class 321, 347, or 348, condition A; and shall have any suitable finish, all in accordance with QQ-S-766C.

(4) *Bolts, nuts, and washers.*—Bolts, nuts, and washers shall be made of corrosion-resisting steel containing 18 percent chromium and 8 percent nickel, or 17 percent chromium and 9 percent nickel.

c. *Fabrication.*—The rubber waterstops shall be molded or extruded and cured in such a manner that any cross section will be dense, homogeneous, and free from porosity and other imperfections. The following minor surface defects will be acceptable:

C-888M

(1) Lumps and depressions not exceeding 6 millimeters in longest lateral dimensions and 2 millimeters deep with no limit to the frequency of occurrence.

(2) Lumps and depressions between 6 and 12 millimeters in longest lateral dimension and 2 millimeters deep as long as the frequency of occurrence does not exceed six in a 15-meter length, and there is at least 50 millimeters between any two such defects.

(3) Marks resulting from the tubing operation or handling during manufacture with no limit to width or frequency of occurrence as long as the thickness of material below the mark is not less than the minimum thickness.

(4) Coarse or grainy surface texture.

(5) Suck-back along flash lines of molded goods if not more than 2 millimeters wide, 2 millimeters deep, and not more than 600 millimeters long.

The tolerances, shown on the drawing, shall govern all cross-sectional dimensions. Any defects which are not within the above limitations either shall be repaired as approved by the contracting officer or shall be removed from the finished product by cutting out a length of waterstop containing such defects and splicing the waterstop at that point. All factory splices shall be molded splices. Molded splices shall be made by vulcanizing the splices in a steel mold for a time sufficient to produce maximum strength in the splice. All molded splices shall withstand being bent 180° around a 50-millimeter diameter pin without any separation at the splice.

[1] (Rubber waterstops for joints in the barrel or box portions of siphons, culverts, or other pressure conduits shall be furnished in continuous circular hoops. The hoops shall be fabricated from straight strip waterstop with the ends spliced together with molded splices to form a closed ring of the specified circumferential length. A tolerance of plus or minus 50 millimeters will be permitted in the circumferential length of each hoop.)

CHAPTER 4—USBR ENGINEERING APPLICATIONS

[1] [d. *Special waterstop intersections.*—Special waterstop intersections, types 1 through _____ as shown on drawing No._____ (_____) shall be fabricated by lap splicing.

The lap splices shall be made by removing the side bulbs and center bulb from the intersecting pieces over the entire overlapping area flush with the webs to provide flat contact surfaces. The open half of the center bulb shall be plugged to form a watertight lap. The contact surfaces of the two overlapping pieces shall then be buffed smooth and coated with rubber cement. A piece of uncured gum rubber shall be applied to one of the contact surfaces, and the two pieces of waterstop shall then be placed together in the position shown on the drawings and be vulcanized in open steam by the best approved methods. During the vulcanizing period, the contact surfaces shall be held firmly together by placing metal plates on both sides of the waterstop and applying C clamps to the metal plates. Suitable gum rubber and rubber cement shall be applied to the butt ends of waterstop at these joints to assure obtaining a continuous watertight seal around the intersection. The lap splice shall withstand being bent $180°$ around a 150-millimeter diameter pin without any separation at the splices.]

_____. *Tests.*—

(1) *General.*—Rubber waterstops shall be subject to laboratory tests before shipment. Material for tests shall be furnished by the manufacturer and all tests shall be made at the place of the manufacturer of the rubber waterstops.

Except as otherwise provided below, general sampling procedures shall be in accordance with section 6 of Federal Test Method Standard No. 601.

(2) *Sampling for tests.*—Samples for laboratory tests to determine physical properties of the compound shall be taken at random to obtain the following number of test units from each separate purchase order:

Size of purchase order	No. of test units
150 meters or less	1
151 to 300 meters	2
301 to 1500 meters	4
1501 to 3000 meters	8
Over 3000 meters	15

At the option of the manufacturer, laboratory tests to determine physical properties of the rubber waterstops required to be furnished under these specifications shall be performed on test specimens cut from (a) test units taken from the finished rubber product or, (b) substitute samples furnished in accordance with paragraph 3.5 of section 6, Federal Test Method Standard No. 601.

(3) *Methods of tests.*—Tests shall be made in accordance with the methods specified in subparagraph b.(1) above.

(4) *Data to be furnished by manufacturer.*—One certified copy of all laboratory test reports representing each shipment of waterstop shall be mailed to the [1] (Project) Construction Engineer, Bureau of Reclamation, _____

_____

_____. *Shipping and storing.*—Rubber waterstops may be shipped in rolls to facilitate handling, but if any roll of waterstop is not to be installed in a structure within 6 months after receipt of the material, the roll shall be loosened. All waterstops shall be stored in as cool a place as practicable, preferably at 21 °C or less. Waterstops shall not be stored in the open or where they will be exposed to the direct rays of the sun.

CHAPTER 4—USBR ENGINEERING APPLICATIONS 111

___. *Installation.*—

(1) General.—The waterstops shall be installed with approximately one-half of the width of the material embedded in the concrete on each side of the joint. Care shall be exercised in placing and vibrating the concrete about the waterstops to insure complete filling of the concrete forms under and about the waterstops and to obtain a continuous bond between the concrete and the waterstops at all points around the periphery of the waterstops. In the event the waterstop is installed in the concrete on one side of a joint more than 1 month prior to the scheduled date of placing the concrete on the other side of the joint, the exposed waterstop shall be covered or shaded to protect it from the direct rays of the sun during the exposure.

[2](____). Field splices in type "A," "B," "G," and "H" rubber waterstops.—All field splices in type "A," "B," "G," and "H" rubber waterstops shall be molded splices. All molded splices shall be made by vulcanizing the splices in a steel mold as follows: The adjoining ends at splices shall be beveled at an angle of $45°$ or flatter by the use of a saw and miter box so that the ends to be spliced together will be pressed together when the mold is closed. The beveled ends and the sides for at least 6 millimeters back from the ends shall be buffed thoroughly to provide clean, rough surfaces. All buffed surfaces shall be given two thin coats of rubber cement and each coat shall be permitted to dry thoroughly. A piece of gum rubber cut to the same dimensions as the beveled face shall then be applied to the end of one strip after removing the cloth backing from the gum rubber. The adjoining strip shall then be placed accurately in position, and all edges shall be stitched thoroughly together with a suitable handstitcher. The mold shall be heated to a temperature of $143\,°C$ before the splice is placed in the mold. The prepared splice shall be placed in the mold with the splice in the center of the mold, and the mold shall be closed tightly to prevent slipping during the vulcanizing process. The splice shall remain in the mold 25 minutes after the mold is closed completely, during which time the mold shall be maintained at a temperature of $143\,°C$.

C-888M

[2] Delete as applicable for splices.

Each finished splice shall withstand a bend test by bending the waterstop 180° around a 50-millimeter diameter pin without showing any separation at the splice.

The contractor shall furnish all materials for splices and all field splicing molds.

[2] (_____). *Field splices in type "D," "E," and "F" rubber waterstops.*—All field splices and intersections in type "D," "E," and "F" rubber waterstops, including connections to metal seals, shall be made as shown on drawing No. _____ (40-D-4567).

_____. *Measurement and payment.*—Measurement, for payment, of the various types of rubber waterstops will be made of the number of metres of waterstops in place measured along the centerline of the waterstop, [2] (with no allowance for lap at splices and intersections.)

Payment for furnishing and placing the various types of waterstops will be made at the applicable unit price per metre bid therefor in the schedule, which unit price shall include the cost of furnishing all material, making field splices and intersections, [2] (making connections,) (making connections to metal seals,) and installing the waterstops.

6-1-76 Revisions: Paragraph revised throughout.

Table 4-4.—*Actual customary values for underlined SI metric values for rubber waterstops specification paragraph (C-888M). These are listed in order of appearance.*

Proposed metric	Actual customary
24.1 MPa	3500 lb/in^2
20.7 MPa	3000 lb/in^2
10 MPa	1450 lb/in^2
7.9 MPa	7900 lb/in^2
mass	weight
0.5 mg/L	0.5 p/m
13.8 MPa	2000 lb/in^2
6 mm	1/4 inch
2 mm	1/16 inch
6 mm	1/4 inch
12 mm	1/2 inch
2 mm	3/32 inch
50 mm	2 inches
2 mm	1/16 inch
2 mm	1/16 inch
600 mm	2 feet
50 mm	2 inches
150-mm	6-inch
6 mm	one-fourth of an inch
50-mm	2-inch

_____. INSULATED GYPSUM WALLBOARD SYSTEM

a. *General.*—The contractor shall furnish all materials and perform all work required for installing the rigid installation and gypsum wallboard on the walls and ceiling of the _____ where shown on the drawings.

b. *Materials.*—

(1) Rigid insulation.—Rigid insulation shall be 40-millimeters thick extruded polystyrene insulation board with dense skin faces on both sides of the boards; shall have a density of not less than 16 kilograms per cubic meter; a thermal conductivity not greater than 35 milliwatts per meter kelvin at 24 °C mean temperature; and a maximum water vapor transmission rate of 2.2×10^{-8} kilogram per pascal second meter and shall be a polystyrene foam board equal to Zonolite Thermo-Stud Board insulation as manufactured by W. R. Grace and Co., Construction Products Division, 62 Whittemore Avenue, Cambridge, Massachusetts 02140.

(2) Steel furring channels.—The steel furring channels shall be No. 25-gage, galvanized, serrated furring channels 40 millimeters wide with 15-millimeter serrated legs and shall be equal to Thermo-Stud serrated furring channels as manufactured by W. R. Grace and Co.

(3) Anchor fasteners.—Anchor fasteners shall be Tapcon flathead concrete anchor fasteners as manufactured by Buildex Division of Illinois Tool Works, Inc., 2500 Brickvale Drive, Elk Grove Village, Illinois 60007, or equivalent.

(4) Gypsum wallboard.—Gypsum wallboard shall be 12-millimeter thick gypsum wallboard in accordance with Federal Specification SS-L-30D, type III, grade R, class L1, form a, style 3.

___M

CHAPTER 4—USBR ENGINEERING APPLICATIONS

(5) *Screws.*—Screws shall be type S or S 12, Bugle Head, screws as manufactured by US Gypsum Co., 101 South Wacker Drive, Chicago, Illinois 60606, or National Gypsum Co., 325 Delaware Avenue, Buffalo, New York 14202.

(6) *Metal trim.*—Metal trim shall be made of galvanized steel or other noncorrosive metal, and shall be of standard commercial quality. Metal trim shall be square type.

(7) *Perforated tape and joint cement.*—Perforated tape and joint cement shall be standard commercial quality.

c. *Installation.*—

(1) *Rigid insulation.*—The rigid insulation boards shall be installed on the interior walls and ceiling where shown on the drawings. The insulation shall be fastened to the concrete by use of serrated channels and anchor fasteners. The serrated furring channels shall be pressed into the insulation boards on 600-millimeter centers with the back of the channel nearly flush with the face of the insulation boards.

The channels shall be attached to the concrete with anchor fasteners spaced at intervals of not more than 450 millimeters in accordance with the recommendations of the manufacturer of the serrated channels. Additional serrated channels shall be installed at corners and at door and other openings.

(2) *Gypsum wallboard.*—The gypsum wallboard shall not be applied to the walls and ceiling until all pipes, electrical conduit, and other features to be concealed are placed. Temperature shall be maintained at 13 to 21 $^{\circ}$C during installation of the wallboard and joint finishing.

All gypsum wallboard shall be applied with the long dimension normal to the furring channels. Wallboard shall be butted together tightly at joints. End joints of wallboard shall be staggered. All ends of wallboard shall bear on a support.

Metal trim shall be installed at locations, as required, to conceal edges of gypsum wallboard. The metal trim shall be installed in accordance with the manufacturer's instructions.

The gypsum wallboard shall be fastened to each furring channel with screw fasteners. The screw fasteners shall be spaced at a maximum of 300 millimeters on centers in the field of the board and at 200 millimeters on centers staggered along abutting ends. All screw fasteners shall be driven with an electric screwdriver of a type recommended by the manufacturer of screw fasteners, and all screwheads shall be driven so as to provide a slight depression below the surface of the board. Screw fasteners shall not be driven closer than 10 millimeters from the edges and ends of the board, and edges of the wallboard and ceiling shall have at least 15 millimeters bearing on all supports.

Joints in exposed gypsum wallboard partition and ceiling facings shall be filled evenly and fully with joint cement using a 100-millimeter putty knife. The cement shall extend approximately 40 millimeters on each side of the joint. Perforated tape shall then be applied directly over the cement. The tape shall be pressed into place with a putty knife, and the cement that is forced through the perforations shall be smoothed down. After the cement has dried thoroughly, another thin coat of cement shall be applied over the tape and shall be feathered out to a width of approximately 100 millimeters on each side of the joint. When the second coat of cement has dried, a third coat of cement shall be applied over the second coat and shall be feathered out to a minimum of 150 millimeters on each side of the joint. Between applications of cement at joints, rough spots or areas shall be sanded smooth. Machine application of perforated tape will be permitted. Twenty-four hours after the third coat has been applied, the entire area covered with cement shall be sanded smooth and level with the wallboard.

___M

All dimples at screwheads and other depressions in the wallboard surfaces shall receive three coats of joint cement applied as each coat is applied to the joints.

After installation, all exposed surfaces of gypsum wallboard and metal trim shall be painted in accordance with paragraph_____(Painting).

d. *Measurement and payment.*—Measurement, for payment, of the insulated gypsum wallboard system will be made of the surface in place. Payment for furnishing and erecting insulated gypsum wallboard system will be at the unit price per square meter bid, therefor in the schedule, which unit price shall include all materials, installation, finishing, and all work required to complete the work.

Table 4-5.—*Actual customary values for underlined SI metric values for gypsum wallboard specification paragraph. These are listed in order of appearance.*

Proposed metric	Actual customary
40 mm	1-1/2 inches
16 kg/m^3	1 lb/ft^3
35 mW/(m·K)	0.24 "K" value, thermal conductivity, BTU-in/(H·ft^2·°F)
2.2 x 10^{-8} kg/(Pa·s·m)	1.5 perm-inch
40 mm	1-5/8 inches
15 mm	5/8 inch
12 mm	1/2 inch
100 mm	4 inches

Materials

Electrical Conductors

As previously stated, the Bureau of Reclamation will purchase materials in units and dimensions competitively available from industry; this includes electrical wire. Table 4-6 lists a limited but representative listing of common conductor sizes showing both the customary and metric dimensions. The metric values are soft conversions and are not presented to indicate possible industry replacement values. Until a new industry standard is established, use the customary units in all procurement specifications and contracts. Metric wire sizes may be developed based upon a rationalized numerical series; see *Rationalized Metric Values* in chapter V. The development of such a series is not considered imminent because of the market position of U.S. manufacturers; any such series may only be a series of soft converted values.

Metric Screw Threads

Metric screw thread dimensions are based upon the ISO basic profile (fig. 4-1); this profile is essentially the same as the Unified screw thread basic form. This profile does not represent actual thread design but serves as the reference to which deviations are applied to form internal and external threads.

Metric screw threads are designated by the capital letter M followed by the basic major diameter (nominal size in millimeters) and the pitch (in millimeters). Note these dimensions in figure 4-1; pitch is represented by the capital P. This method of expressing pitch is quite different from that used in the Unified system. In the Unified system, pitch is expressed as threads per inch; in the metric system pitch is the distance between the screw crests. For coarse series threads, the indication of pitch is omitted.

CHAPTER 4—USBR ENGINEERING APPLICATIONS

Examples: Coarse series threads   - M8
                                    M6
           Other threads   - M8 x 1
                                    M6 x 0.75

A complete designation of an ISO metric screw thread also includes an identification of the tolerance class. The tolerance class consists of a tolerance grade and a tolerance position for the crest diameter and the pitch diameter: when these two tolerance symbols are identical, the symbol is only shown once and not repeated. See figure 4-2 for examples of complete ISO metric screw thread designations.

The number of a tolerance grade reflects the size of the tolerance; grade 8 tolerances are greater than grade 6 which are greater than grade 4. Refer to

Table 4-6.—*Equivalent metric dimensions of standard electrical conductor sizes* [1]

Customary data			Metric data	
Thousand circular mils	AWG No.	Diameter[2] in	Diameter[3] mm	Cross-sectional area, mm^2
—	6	0.198	5.03	13.30
—	4 (6/1)	0.250	6.35	21.15
—	2 (6/1)	0.316	8.03	33.63
—	1	0.355	9.02	42.41
—	0	0.398	10.11	53.46
—	2/0	0.447	11.35	67.44
—	3/0	0.502	12.75	85.03
—	4/0	0.563	14.30	107.22
80.0	—	0.367	9.32	40.54
101.8	—	0.461	11.71	51.58
110.8	—	0.481	12.22	56.14
134.6	—	0.530	13.46	68.20
159.0	—	0.576	14.63	80.57
176.9	—	0.607	15.42	89.64
190.8	—	0.663	16.84	96.68
211.3	—	0.609	15.47	107.07
266.8	—	0.642	16.31	135.19
300.0	—	0.684	17.37	152.01
336.4	—	0.721	18.31	170.46
397.5 (18/1)	—	0.783	19.89	201.42
477.0 (18/1)	—	0.846	21.49	241.70
556.5 (24/7)	—	0.914	23.22	281.98
615.0 (24/7)	—	0.953	24.21	311.63
636.0 (18/1)	—	0.940	23.88	322.26
666.0	—	1.000	25.40	337.47

[1] IEEE/ICC working group 934 is working on hard metrication of conductor tables.
[2] To convert from inches to millimeters, multiply by 25.4.
[3] To convert from thousand circular mils (kcmil) to square millimeters (mm^2), multiply by 0.506 707.

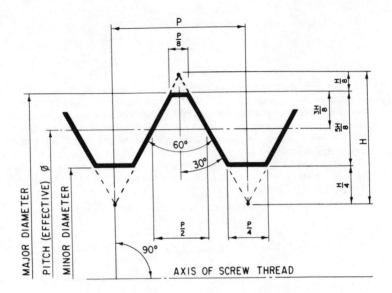

Figure 4-1.—ISO basic thread profile. 40-D-6338

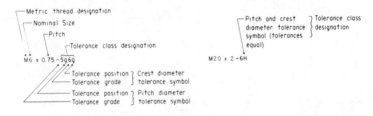

Figure 4-2.—Typical thread and tolerance designations. 40-D-6339

table 4-7 for a listing of the tolerance grades and the dimensions to which they may be applied. Grade 6 tolerances are normally used for medium quality and normal length of engagement applications; smaller tolerance grades are used for fine quality and/or short lengths of engagement. Conversely, tolerance grades above grade 6 are used for coarse quality and/or long lengths of engagement. The pitch diameter tolerance is not to exceed the crest diameter tolerance; this means that for some tolerance grades, certain tolerance values for fine pitches are not applicable due to insufficient thread overlap.

Length of engagement is the actual length of assembled thread mating with the corresponding part. This is almost always less than the full thread length; it cannot be greater. Length of engagement is classified as short (S), normal (N), or long (L). The length of engagement symbol may be added to the basic thread designation. Examples of this are: M6-7g6g L and M8 x 0.75-4g S.

Tolerance positions define the maximum material limits of the pitch and crest diameters relative to the basic profile. Table 4-8 shows position symbols applicable to internal and external threads.

Table 4-7.—*Tolerance grades*

External thread		Internal thread	
Major diameter	Pitch diameter	Pitch diameter	Internal thread
—	3	—	—
4	4	4	4
—	5	5	5
6	6	6	6
—	7	7	7
8	8	8	8
—	9	—	—

Table 4-8.—*Tolerance position symbols*

External threads	e[1]	large allowance
	g	small allowance
	h	no allowance
Internal threads	G	small allowance
	H	no allowance

[1] Not to be applied to pitches finer than 0.5 mm.

The basic thread designations may also include information relating to thread fits, rounded root threads, and designation of coated threads. Refer to the ASME B1 Report [3] for further information regarding this and other data.

Listed in the following two tables (tables 4-9 and 4-10) are the standard series threads to be used for the production of commerical screw thread products and the preferred tolerance classes.

The major U.S. fastener manufacturers, when it appeared the U.S. was going metric, took the opportunity to examine the existing ISO system of metric fasteners. Possible technical improvements or economic benefits were examined, and the result was the OMFS (Optimum Metric Fastener System).

The OMFS was based upon a more simple basic profile (fig. 4-3) which allowed for greater fatigue resistance. The simplified basic thread form does not have a pitch diameter but it does have a flank diameter tolerance equivalent to the ISO class 5g pitch diameter tolerance. The thread depth is slightly shallower with a larger radius at the root.

Table 4-9.—*Metric threads for commercial screws, bolts, and nuts*

Nominal size diameter (mm)		Pitch (mm)	
[1]Column 1	Column 2	Coarse thread	Fine thread
0.25		0.075	
0.3		0.08	—
	0.35	0.09	—
0.4		0.1	—
	0.45	0.1	—
			—
0.5		0.125	—
	0.55	0.125	—
0.6		0.15	—
	0.7	0.175	—
0.8		0.2	—
	0.9	0.225	—
1		0.25	—
	1.1	0.25	—
1.2		0.25	—
	1.4	0.3	—
1.6		0.35	—
	1.8	0.35	—
2		0.4	—
	2.2	0.45	—
2.5		0.45	—
3		0.5	—
	3.5	0.6	—
4		0.7	—
	4.5	0.75	—
5		0.8	—
6		1	—
	7	1	—
8		1.25	1
10		1.5	1.25
12		1.75	1.25
	14	2	1.5
16		2	1.5
	18	2.5	1.5
20		2.5	1.5
	22	2.5	1.5
24		3	2
	27	3	2
30		3.5	2
	33	3.5	2
36		4	3
	39	4	3

[1] Thread diameter should be selected from column 1 or 2, with preference being given in that order.

Table 4-10.—Preferred tolerance classes

Quality	External threads (bolts)											Internal threads (bolts)						
	Tolerance position e (large allowance)			Tolerance position g (small allowance)			Tolerance position h (no allowance)					Tolerance position G (small allowance)			Tolerance position H (no allowance)		
	Length of engagement			Length of engagement			Length of engagement					Length of engagement			Length of engagement		
	Group S	Group N	Group L	Group S	Group N	Group L	Group S	Group N	Group L			Group S	Group N	Group L	Group S	Group N	Group L
Fine							3h4h	4h	5h4h						4H	5H	6H
Medium		6e	7e6e	5g6g	6g	7g6g	5h6h	6h	7h6h			5G	6G	7G	5H	6H	7H
Coarse					8g	9g8g							7G	8G		7H	8H

OMFS

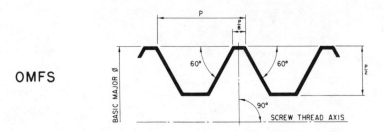

Figure 4-3.—OMFS basic thread profile. 40-D-6340

An additional benefit of the OMFS was the reduced number of diameter/pitch combinations; it is less than half the ISO number for corresponding screw thread sizes. Other benefits attributed to the OMFS are: (1) there is only one thread series, (2) interchangeability with ISO metric fasteners, and (3) new boundary profiles for inspection gaging. The OMFS was standardized by the IFI (Industrial Fasteners Institute) and described in their standard IFI-500, *Screw Threads for Metric Series Mechanical Fasteners,* published in 1974.

The ANSI Committee B1 and the ISO Technical Committees 1 and 2 have reached an agreement to include certain elements of IFI-500 into ISO standards 68, 261, and 965. This agreement awaits final ratification by ISO members but approval is considered certain. A U.S. specification is being written by ANSI in anticipation of a ratified agreement. The OMFS designation will be dropped. The extent to which the general characteristics of the OMFS will be incorporated into the ISO standards is not known at this time.

This will not affect metric threaded hardware presently on the market. No interchangeability problems are anticipated for new hardware which will be manufactured to the new industry standard being prepared by ANSI.

Pipe Dimensions

The AWWA (American Water Works Association) has proposed nominal SI metric values to be used to describe various pipe sizes now being used. Although there are no known pipe sizes being manufactured to these metric dimensions, this should not create any problems; the use of nominal metric values has its precedent in the U.S. customary values used to describe pipe. There is no 1-inch dimension associated with 1-inch pipe; therefore, describing the same pipe as 25-mm pipe will not result in a loss of precision.

As and when established SI metric standards are prepared, these standards will then serve as the pipe size designations. Ideally, the new standards will use a rationalized size series like those described in chapter V.

Table 4-11 lists the suggested nominal SI metric pipe sizes; during the interim period of soft conversions, these millimeter designations will be an acceptable substitute if their use is desired and recognized by commercial suppliers.

Table 4-11.—*Nominal SI metric pipe sizes*

Sizes less than 4 in		Sizes equal to or greater than 4 in	
U.S. customary (inch)	Metric (mm)	U.S. customary (inch)	Metric (mm)
1/4	8	4	100
3/8	10	6	150
1/2	15	8	200
5/8	16	10	250
3/4	20	12	300
1	25	16	400
1-1/4	32	20	500
1-1/2	40	24	600
2	50	30	750
2-1/2	65	36	900
3	80	42	1050
3-1/2	90	48	1200
		See footnotes 1, 2, and 3	

[1] The succeeding and intermediate sizes not shown may be determined by the nominal relation of 1 inch = 25 millimeters.

[2] For very large diameters, convert directly (1 in = 25.4) and then round to the nearest multiple of 25. For example, a 120-inch diameter should be described as a 3050-mm diameter instead of a 3000-mm diameter.

[3] The recommended equivalents are in accordance with British and German standards.

In addition to the nominal pipe sizes, the AWWA has recommended other units for selected quantities; these are listed in table 4-12.

Guidelines for SI Design Drawings

Introduction

The following drafting guidelines have been developed for use by the Bureau of Reclamation, based upon ISO standard practices and internal procedures. These are interim guidelines and are not to be considered final until a new drafting standard is developed to reflect ISO practices for SI design drawings.

Format

Until the ISO-size paper is commercially and competitively available for use, all drawings will be prepared on the regulation size paper now in use. Information regarding the available international size paper is presented in a succeeding section.

Table 4-12.—*AWWA recommended units for pipe dimensions**

Feature	Unit of measure
pipe diameter	millimeters
wall thickness	millimeters
length	meters
mass per length	kilogram per meter
test pressure	kilopascals
pipe strength	kilopascals
pipe threads	no recommendation

* Refer to USBR preferred units tables (chapter III) for units to be used by the Bureau.

All SI metric drawings will display the USBR SI metric identification symbol as shown on figure 4-4. This symbol shall appear as shown on figure 4-5. On standard 21- by 36-inch drawings, the symbol is to be 10 mm tall; use a 5-mm-tall symbol on report size sheets (8 by 10-1/2 and 8-1/2 by 11 inch) and legal size sheets.

Multiview projections.—First and third angle projections are simply common means of drawing a three-dimensional object and displaying all pertinent views. In the United States and Canada, the third angle projection is the traditional scheme used in multiprojections. The drafting SI/ISO practice of using third angle projection will not change the Bureau's present method of drawing preparation and general layout.

The ISO recommends that the first angle projection symbol or the third angle projection symbol be used on design drawings when either one of these layout formats is used. The Bureau's preferred method is the use of viewing plane or cutting plane lines to indicate views and sectional views. When this method is used, the third angle projection symbol need not be shown since this method is self-explanatory. However, on design drawings that use the third angle projection format and the section lines are not employed, the third angle projection symbol should be shown. The symbol is to appear as shown in figure 4-5. On standard 21- by 36-inch drawings, the symbol is to be 10 mm tall; use a 5-mm-tall symbol on 8- by 10-1/2- and 8-1/2- by 11-inch sheets.

Refer to figure 4-6 for a graphic example of the difference between the first and third angle projections.

Units for Design Drawings

The millimeter will be the basic dimensioning unit; other SI units of linear measure to be used on drawings include the kilometer, meter, and micrometer. The use of U.S. customary units will also be permitted as required by design or product standards.

The Bureau's design drawings can be classified into four categories: civil, structural, mechanical, and electrical. Table 4-13 lists the units to be used for drawings falling into the structural and civil categories; example drawings are also identified.

CHAPTER 4—USBR ENGINEERING APPLICATIONS

Figure 4-4.—USBR SI metric symbol. 40-D-6341

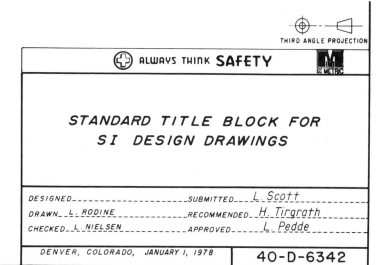

Figure 4-5.—USBR standard title block. 40-D-6342

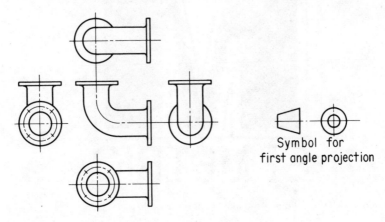

Pipe bend, first angle projection – European style

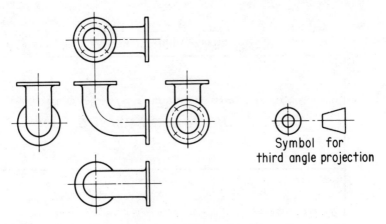

Pipe bend, third angle projection – U.S. / Canadian style

Figure 4-6.–Comparison of multiview projection methods and symbols. 40-D-6343

CHAPTER 4—USBR ENGINEERING APPLICATIONS

Table 4-13.—*Dimensional units for design drawings*

	Structural drawings	Civil drawings
Linear Elevations[1] Stations[1]	millimeters meters, 3 decimal places meters	meters meters, 2 decimal places meters
Example drawings[2]	powerplant pumping plants control houses valve houses outlet houses tunnels bridges buildings	site plans location plans yard layouts excavation drawings stilling basins canals large spillway and outlet works dams: concrete or earth

[1] A note must be included on the appropriate drawing stating "Stations and elevations are in meters." This is required since the meter symbol (m) will not follow the numerical values. (Note: one station equals 100 m.)

[2] The drawing list does not attempt to show all the possible project types, but should be used as a guide for selecting the appropriate metric units. Also, the size of the facilities listed will have an influence on the metric units used on the drawings.

Survey angles will continue to be shown in degrees, minutes, and seconds, or in degrees and decimals of a degree, rounded to the nearest thousandth. There is not an anticipated change to surveying equipment which will measure bearing angles in radians; therefore, the present practice of using angular measure will not change.

Mechanical drawings will be dimensioned in millimeters except for surface textures, elevations, and stations. Electrical drawings will basically be dimensioned in millimeters **except** where there will be an interface with large scale civil drawings. For either type of drawing, do not use decimal fractions of a millimeter unless such precision is absolutely required.

With few exceptions, the SI units of linear measurements are not to be intermixed on the same drawing; for example, do not show distance between column lines in meters and locate equipment in millimeters. The specified unit of linear measurement should be used throughout the drawing; this rule does not affect elevations, station dimensions, and surface textures.

Drafting Practices

Material designation.—Material designations for standard or nominal U.S. customary dimensions are not to be soft converted. The customary designations are to be retained until standard or nominal metric sizes are commercially available. Fabricated material or material with nonstandard or nominal dimensions will be expressed in SI units.

It is recognized that hybrid drawings will result until standard and nominal size equipment is metricated. The intermixing of metric and customary units on drawings will be accepted at this time. Examples are as follows:
1. PL 1/2" x 300 x 562
2. (8X) φ 1/2" ↧ 12 ± 0.3
3. L 3" x 2" x 1/4" x 500
4. φ 2" Conduit x 3000
5. φ 1/2" Reducing elbow[1]

The dimension values which show no unit symbols are to be assumed in the basic unit of the drawing, e.g., millimeters.

The designer should follow the USBR standard practices required for surface texture, tolerance symbols, etc. Using SI units for such designations must be preceded by actual conversion of these standards to metric units and symbols. The weld standard (AWS A2.3-75) has been converted and the use of SI units for weld callouts is now required. The surface texture specification (ANSI B46.1) has not been metricated; therefore, the use of microinches will continue. See ISO 1302 for the method of indicating surface textures on drawings.

Items covered by standards not converted to reflect SI metric should continue to be designated using U.S. customary units and symbols. Whenever a non-SI unit is used on a drawing, clearly indicate the unit of measure.

Dual dimensioning is the showing of SI metric units together with the U.S. customary units in parentheses, for example, 100 mm (3.94 in). **Dual dimensioning on drawings is discouraged**, except where the dimensions must interface with existing installations or construction in U.S. customary units.

When interfacing a metric design to a customary unit design, the metric and customary unit equivalents should be shown in a table appropriately placed on the drawing; all dimensions on the drawing should be shown in the table. Whenever it is necessary to show metric and U.S. customary dimensions on a drawing, the designer shall use the format shown on figure 4-7.

When dual dimensioning is used on a drawing, the parts shown on the drawing must achieve the required degree of interchangeability regardless of which system of measure, SI or customary, is the one used to manufacture or construct the parts. Interchangeability is determined at the time of dimensional conversion by the number of decimal places retained when rounding a converted dimension. Refer to chapter V for information concerning rounding procedures required for material fits and interchangeability.

Expressing dimensions.—Refer to chapter II for complete information concerning style and usage of SI unit names and symbols. Particular elements applicable to design drawings are repeated here.

In accordance with ISO drafting practices, numerical values and letters for SI unit prefixes and names (or symbols) are to be printed using vertical letters. The Bureau currently uses letter guides with slanted letters. The use of these guides will be permitted, even for numerical values and SI names and symbols. While lettering guides with upright letters are available and are preferred for numerical values and symbols, their use is not meant to exclude the slanted lettering. The use of vertical lettering may be applied to the numerals and unit

---

[1] Note there is no letter space left between the numeral and customary unit symbols.

CHAPTER 4—USBR ENGINEERING APPLICATIONS

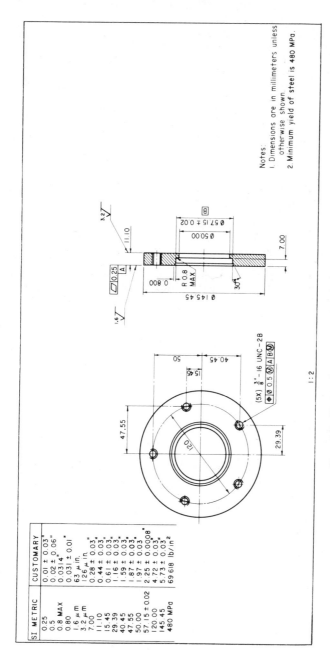

Figure 4-7.—Illustrated use of dual dimensioning table. 40-D-6344

names (or symbols) while the rest of the lettering is completed with the slanted lettering. However, drawings must be consistent in the type of lettering used for numerals and SI symbols; do not intermix slanted and vertical lettering for these drawing elements.

Always leave a space between a numerical value and an SI unit name or symbol. This also includes numerical values for temperatures such as 25 °C and 300 K. This does not apply to non-SI unit symbols such as feet ('), inches ("), degrees (°), and so forth.

Angular dimensions should be expressed in degrees, minutes, and seconds, such as 22°, 22°30', and 22°30'15", or in decimals of degree, such as 35.025°. A full letter space should be used to write the degree, minute, and second symbols leaving no space between the symbol and the preceding and following numerical values. When an angle is less than 1°, a preceding zero must be used, for example, 0°15', 0°0'30", or 0.750°.

Commas should not be used to denote thousands with either U.S. customary or SI metric units because it is the practice in many countries to use the comma as a decimal marker. On engineering drawings, numerical values should be grouped in three digit groups counting to the right and left of the decimal marker. Numbers having four digits to the right or left of the decimal marker are to be blocked together leaving no space where the comma would normally be used. Numbers having five or more digits to the right or left of the decimal marker should have a space where the comma would normally be used to denote thousandths.

Examples: [2][3]STA 123+59.986   STA 0+78.123
EL 3432.063   EL 10 051.00
28 432 mm   8432 mm
32.063 m

Always space a number having four or more digits to the right or left of a decimal marker when it appears in a table of numerical values.

The decimal marker used with metric units is a period (full stop symbol) which should be **bold**, given a full letter space, and be placed on the baseline, as 25.4.

For any dimension less than 1.0, a zero must precede the decimal marker (e.g., 0.12 m).

Do not add insignificant zeros to the right of the decimal point, except as required for necessary precision. If a millimeter dimension is a whole number, the decimal marker and zeros are not to be shown to the right of the numerical value.

---

[2] The symbol for station (STA) and elevation (EL) consists of capital, vertical letters; a period is not used.

[3] For station designations, the grouping of numbers into groups of three digits will not normally occur. No more than three decimal positions will ever be given, so the only question concerns the digits to the left of the decimal marker. For stations only, treat the plus sign (+) as if it were a second decimal marker; that is, count the digits to the left of the plus sign to determine the grouping format. Examples of this procedure are STA 1234+56.789 and STA 12 345+67.890.

Where equidistant or regularly arranged elements appear on a drawing, the following is used for simplicity:

Use (3X) 508 = 1524 in place of three spaces @ 508 = 1524.

It is preferred that dimensions be written in the length-width-height sequence. The numerical values are separated simply by a multiplication sign (x); a space is left on both sides of the multiplication sign.

    Examples: Notes: 9.5- x 4.8-mm keyway
                                     4- x 3- x 6-m box

                Drawing body: 9.5 x 4.8 keyway
                                       4 x 3 x 6 box

For the standard sequence, it is not necessary to show the units when the designation is part of the drawing; the unit symbol must follow the last number when the designation is part of the notes.

If the dimensions are written in any other sequence which is not established by a design or product standard, each dimension must be followed by the unit symbol and the dimension identification. This format must be followed for notes and callouts listed on the drawing.

    Examples: 1.56-m-deep x 1.00-m-wide trench
                       900-mm-long x 1200-mm-high x 100-mm-wide box
                       PL 1/2" x 300 x 562

The third item in the preceding example does not follow the preferred sequence and does not contain SI unit symbols or any dimension identification; however, this is still a correct designation since it is in the format used by the steel industry for plate steel. The inclusion of the inch symbol is necessary as this will be the Bureau's practice for all non-SI units. Table 4-14 shows other designations for structural steel shapes which do not follow the SI/ISO practices; these were excerpted from the AISC (American Institute of Steel Construction) manual and modified to minimize conflict with SI/ISO practices. This and other industry standards such as ACI 315-74 will be the preferred source of correct material designations.

Tolerances.—Figure 4-8 illustrates the format and relative numeral sizes to be used in presenting the tolerance types described in the following.

When unilateral tolerances are used and either the plus or minus tolerance is nil, a single zero is shown without a plus or minus sign, or a decimal marker. The tolerances are written in a smaller size than the nominal value; the plus tolerance is written one-half line up and the minus tolerance is written one-half line below the nominal value.

When bilateral tolerances are used, both the plus and minus tolerances must have the same number of decimal places; the addition of insignificant zeros is required when necessary to accomplish this. The size and position of the tolerance values are the same as for unilateral tolerances, provided the tolerances are unequal. For equal tolerances, the tolerance value is written using a plus and minus sign (±); the size of the tolerance numerals are placed

Table 4-14.—*Hot-rolled structural steel shape designations*

Designation	Type of shape
W 24" x 76# W 14" x 26#	W shape
S 24" x 100#	S shape
M 8" x 18.5# M 10" x 9# M 8" x 34.3#	M shape
C 12" x 20.7#	American Standard Channel
MC 12" x 45# MC 12" x 10.6#	Miscellaneous Channel
HP 14" x 73#	HP shape
L 6" x 6" x 3/4" L 6" x 4" x 5/8"	Equal Leg Angle Unequal Leg Angle
WT 12" x 38# WT 7" x 13#	Structural Tee cut from W shape
ST 12" x 50#	Structural Tee cut from S shape
MT 4" x 9.25# MT 5" x 4.5# MT 4" x 17.15#	Structural Tee cut from M shape
PL 1/2" x 18"	Plate
Bar 1" ⊡ Bar 1-1/4" φ Bar 2-1/2" x 1/2"	Square Bar Round Bar Flat Bar
Pipe 4" Std. Pipe 4" x—Strong Pipe 4" xx—Strong	Pipe
TS 4" x 4" x .375" TS 5" x 3" x .375" TS 3" OD x .250"	Structural Tubing: Square Structural Tubing: Rectangular Structural Tubing: Circular

CHAPTER 4—USBR ENGINEERING APPLICATIONS

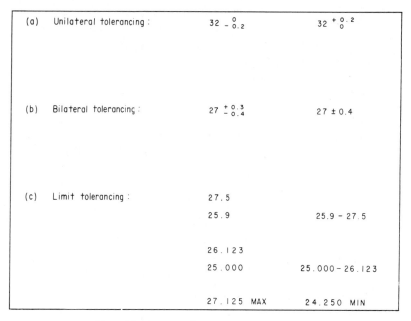

Figure 4-8.—Examples of tolerance formats. 40-D-6345

on the same line with the nominal value and are the same size. The upper deviation is always written above the lower deviation.

For limit tolerances, the high limit is placed above the low limit. Never separate the high and low limits by a line. Both limits are expressed to the same number of decimal places using additional zeros when necessary. An alternative method to the stacked method is the single line expression. For a single line expression, the low limit precedes the high limit and they are separated by a dash; again, the number of decimal places must be equal.

For dimensions limited in one direction only, add MAX or MIN as appropriate. An example of this is shown in figure 4-8.

The use of tolerance indicators to designate approximate dimensions (e.g., 100 ±) has long been used by the Bureau; this consists of giving the nominal value followed by a plus and minus sign but with no tolerance value. This does not indicate plus and minus zero, but does indicate an approximate dimension indicating sufficient latitude to "make fit." This practice is not endorsed, only mentioned for information purposes.

The aforementioned tolerance guidelines also apply to angular measures.

Form and position tolerance symbols are shown in figure 4-9. A form or position tolerance is stated in a feature control symbol shown in figure 4-10. The vertical line separates the symbol from the tolerance. When a tolerance of form or position is related to a datum, the relationship is stated in the feature control symbol by placing the datum reference letter following the tolerance.

Characteristics to be toleranced		Symbols
Form of single features	Straightness	—
	Flatness	▱
	Roundness	○
	Cylindricity	⌀/
	Profile of any line	⌒
	Profile of any surface	⌓
Orientation of related features	Parallelism	//
	Squareness	⊥
	Angularity	∠
Position of related features	Position	⊕
	Concentricity and coaxiality	◎
	Symmetry	≡
Run-out		↗

Figure 4-9.—Form and position tolerance symbols. 40-D-6346

Figure 4-10 shows one of the many ways that form and position tolerances can be indicated. Refer to ISO R1101, ISO R1660, and ISO R1661 for additional information.

Scales and scaling —The ISO recommended scale ratios are in-scale multipliers and reciprocals of 1, 2, 5, and 10; this series is referred to as the rationalized decimal range series. These ratios are a result of rounding the Renard R3 series which is based on the cube root of 10. The resultant representative ratios are listed in table 4-15.

CHAPTER 4—USBR ENGINEERING APPLICATIONS

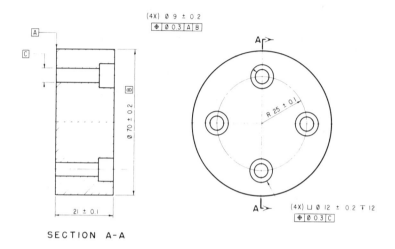

Figure 4-10.—Example of position tolerance symbol. 40-D-6347

Table 4-15.—*ISO standard metric drafting scale ratios*

5:1	1:20	1:2 000	1:200 000
2:1	1:50	1:5 000	1:500 000
1:1	1:100	1:10 000	1:1 000 000
1:2	1:200	1:20 000	1:2 000 000
1:5	1:500	1:50 000	1:5 000 000
1:10	1:1 000	1:100 000	

Comparison of U.S. customary scale ratios to ISO scale ratios is shown in table 4-16. The use of scale ratios other than those listed in table 4-15 should generally be avoided since they are not ISO recommended ratios.

Drawing scales.—All drawings which contain figures shall have bar scales and/or scale ratios, or be labeled "not to scale." Drawings which consist of parts lists, operating instructions, or otherwise contain no figures are not required to indicate any scale. The preferred method to show drawing scale is the use of bar scales located in the lower left-hand corner of the drawing. The use of bar scale(s) provides a convenient method of revising a drawing when the original copy is not available after it has been reduced in size through photocopy methods, or has been placed on microfilm and then enlarged.

Refer to figure 4-11 for the format of bar scales for the scale ratios between 2:1 and 1:5 000 000. Bar scales which show the length measurements in millimeters will not require unit identification; for example, for the 1:100 scale ratio, 1 mm (measured) represents 100 mm (actual), figure 4-11, sheet 1. For the larger scale ratios, it may be desirable to use an alternative unit; for

Table 4-16.—*Customary versus ISO drafting scale ratios*

Customary scale and ratio		ISO ratio
Double size	2:1	2:1
Full size	1:1	1:1
Half size	1:2	1:2
3 in = 1 ft	1:4	1:5
1-1/2 in = 1 ft	1:8	1:10
1 in = 1 ft	1:12	1:10
3/4 in = 1 ft	1:16	1:20
1/2 in = 1 ft	1:24	1:20
1 in = 2 ft	1:24	1:20
3/8 in = 1 ft	1:32	1:20 or 1:50
1 in = 3 ft	1:36	1:20 or 1:50
5/16 in = 1 ft	1:38	1:20 or 1:50
1/4 in = 1 ft	1:48	1:50
1 in = 4 ft	1:48	1:50
1 in = 5 ft	1:60	1:50
3/16 in = 1 ft	1:64	1:50
1 in = 6 ft	1:72	1:50 or 1:100
1/8 in = 1 ft	1:96	1:100
1 in = 8 ft	1:96	1:100
1/16 in = 1 ft	1:192	1:200
1/32 in = 1 ft	1:384	1:200 or 1:500

example, the 1:5 000 000 bar scale may be expressed in meters or kilometers (e.g., 1 mm represents 5000 m, or 1 mm represents 5 km, see figure 4-11, sheet 2). This choice of units provides the opportunity to reduce the size of numbers appearing on the bar scale; the drawing's basic dimensioning unit should provide the key as to whether a unit symbol should appear on the bar scale and the magnitude of the scale numbers. When the millimeter is not the basic unit, add the unit symbol after the numerical unit on the right-hand end of the bar scale.

Figure 4-12, a through e, illustrates the use of bar scales and scale ratios for the more common situations that will occur. Figure 4-12(a) shows a drawing with several details or views but relatively few scale ratios. Each detail shows a scale ratio below each section title that can then be matched to the appropriate bar scale which will have the identical scale ratio below it. The detail which was not drawn to any particular scale is labeled NO SCALE. If the drawing would have required an excessive number of bar scales (e.g., 5 or more) which could not have been accommodated in the available space, it would have been "acceptable" to delete all bar scales and show only the scale ratios under each detail.

Figure 4-12(b) illustrates the layout and format with the greater preference. For this format, the bar scale used for any particular detail is identified with a note under the scale along with the scale ratio identification. The scale ratios

CHAPTER 4—USBR ENGINEERING APPLICATIONS

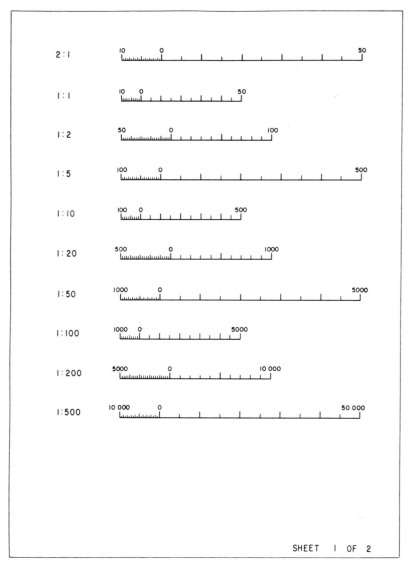

Figure 4-11.—The preferred bar scale graduations for scale ratios between 2:1 and 1:5 000 000. 40-D-6348-1 (2 sheets)

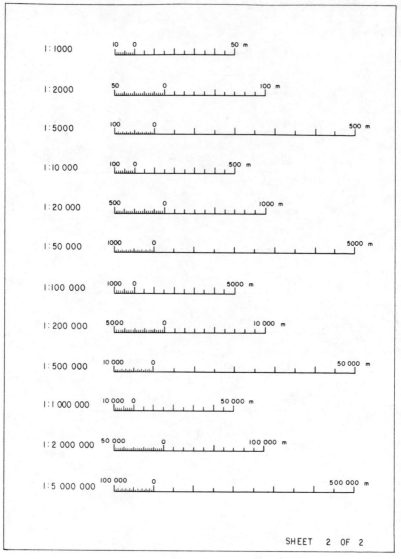

Figure 4-11.—The preferred bar scale graduations for scale ratios between 2:1 and 1:5 000 000. 40-D-6348-2 (2 sheets)

CHAPTER 4—USBR ENGINEERING APPLICATIONS 141

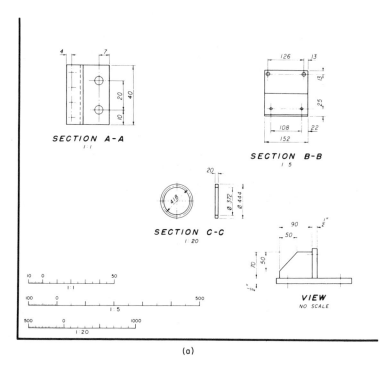

(a)

Figure 4-12.—Use of bar scales on design drawings. 40-D-6349-1 (3 sheets)

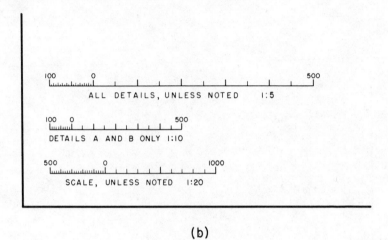

(b)

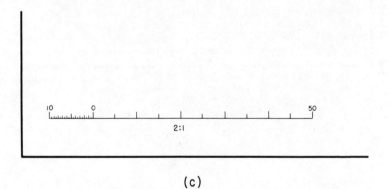

(c)

Figure 4-12.—Use of bar scales on design drawings. 40-D-6349-2 (3 sheets)

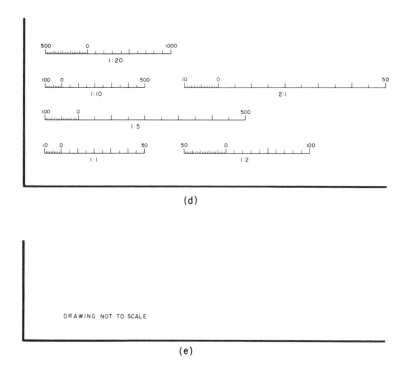

Figure 4-12.—Use of bar scales on design drawings. 40-D-6349-3 (3 sheets)

are not shown under each detail or view. Figure 4-12(c) is the format used when a drawing has only one scale; the scale ratio is not shown anywhere else on the drawing. On this type of drawing, it is still permissible to have views or details which are not drawn to scale provided they are labeled NO SCALE.

Figure 4-12(d) is similar to 4-12(a) except the placement of the scales has been altered to utilize the space available. The departure from the single-column format is acceptable (not preferred) when the inclusion of the bar scales is necessary (i.e., identification of scale ratios only is inadequate) and space limitations do not permit the single-column format. Another alternative (not shown) would be to show the bar scale under the associated detail or view.

Figure 4-12(e) is self-explanatory. A drawing might not be drawn to scale for various reasons, such as when dimensions are changed and redrawing is not warranted, or features may intentionally be drawn out of scale and exaggerated for clarity. In cases such as this, a bar scale need not be added to a drawing or detail. The following should be added in the lower left-hand corner of the drawing: DRAWING NOT TO SCALE. The use of the "drawing not to scale" may also be desirable even if drawings are to scale but the designer does not wish to have unidentified dimensions scaled from the drawing.

Symbols.—Welding sizes and symbols will be in accordance with the SI metric specifications listed in the American Welding Society Structural Welding Code, AWS A2.3-75.

Surface texture (finish) specifications and symbols will remain in microinches as specified by the American National Standard Institute standard B46.1 until the standard is revised for SI units. See ISO 1302 for international method of indicating surface textures (finishes) on drawings.

The abbeviations D., dia., and DIA. for diameter are customary notations and not accepted by ISO. The ISO symbol for diameter is ϕ. A capital R is the symbol for radius. Use □ to indicate a square section. The ϕ, R, and □ symbols precede the numerical values. When it is necessary to indicate inside or outside diameters, spherical diameter, or radius, the dimension should appear as shown on figure 4-13. These and other drawing symbols to be used on SI metric drawings are shown in table 4-17.

For the method of dimensioning holes and the interpretation of the dimensioning, see figure 4-14. For the method of dimensioning and interpretation of tapped holes, counterbore, and countersink, see figures 4-15 and 4-16.

The use of customary abbreviations can be confused with SI symbols. Avoid the use of customary abbreviations and acronyms unless they are clearly defined on the drawing, in the notes, or by a specified technical standard (e.g., ANSI Y1.1). It is preferred to spell out the abbreviation or acronym to reduce the possibility of confusion and misunderstanding.

Tables A-5 and A-6 in the appendix provides a limited list of acceptable abbreviations and symbols which may be used on SI design drawings.

Notes.—When needed, all drawings shall state the appropriate units of dimensions in a note as follows:
1. Dimensions are in millimeters unless otherwise shown.
2. Dimensions are in meters unless otherwise shown.
3. Dimensions are in kilometers unless otherwise shown.
4. Surface textures are in microinches.[4]
5. Contour intervals are in   -meter increments.[5]
6. Elevations are in meters.
7. Stations are in meters.
8. Stations and elevations are in meters.

Other notes will generally consist of instructions not shown on a dimension line. Whenever a numerical value is included in a note, the proper SI unit symbol must be included.

Examples: 1. Make piston clearance in cylinder 0.025 to 0.045 mm.
2. A 9- x 5-mm keyway to be cut for **26** with **22** and **24** assembled.

---

[4] The use of microinches is permitted in lieu of micrometers pending the issuance of new ANSI standards and the availability of gages graduated in micrometers.
[5] Substitute appropriate numerical value; see *Mapping* and *Surveying* section.

CHAPTER 4—USBR ENGINEERING APPLICATIONS

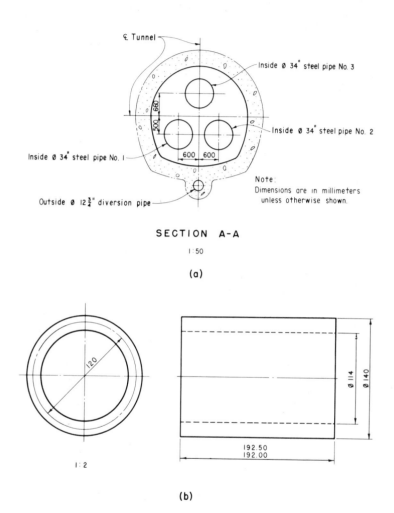

Figure 4-13.—Dimensioning inside and outside diameters. 40-D-6350

Table 4-17.—*ISO drafting symbols*

Symbol	Comment	Example
□	Square section	□ 2 ± 0.2
∅	Diameter	∅ 6 ± 0.4
R	Radius	R 12 ± 0.8
X	Times or places. Used to indicate repetitive features. *	(2X) ∅ 4
⊤	Depth. Replaces the use of depth.	∅ 4 ⊤ 10 ± 1
⊔	Counterbore or spotface	⊔ ∅ 10 ⊤ 8
∨	Countersink	∨ 60° ∅ 9 ± 0.5
I	Thick or thickness	□ 12 I 0.5
△	Chamfer	△ 45° x 6 ± 1
( )	Parentheses. Used to indicate repetitive features.	(7X) ∅ 9
⌀	Tapped hole	See figures 4-15 and 4-16
⌀	Counterbored tapped hole	

* Use a capitalized X to indicate repetitive features; use a lowercase x for the dimensional joiner "by". Do not use to indicate material quantities.

CHAPTER 4—USBR ENGINEERING APPLICATIONS

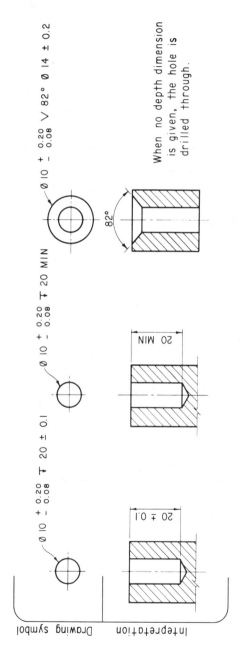

Figure 4-14.—The dimensioning and interpretation of holes. 40-D-6351

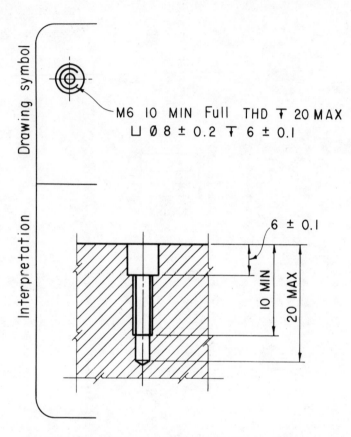

Figure 4-15.—The dimensioning and interpretation of a counterbored, tapped hole. 40-D-6352

CHAPTER 4—USBR ENGINEERING APPLICATIONS

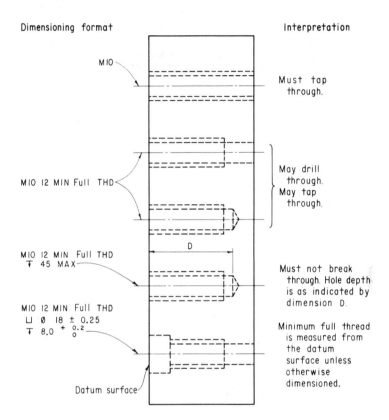

Figure 4-16.—Examples of various tapped hole designations and the interpretation. 40-D-6353

Surveying and Mapping

Elevations—Elevations will be shown in meters. The number of decimal places used will be a function of the precision required. Elevations for maps and concrete and steel work will be shown to three decimal places (e.g., EL 495.200). Elevations for earthwork and excavations will be shown to two decimal places (e.g., EL 495.26).

Stations.—Stations will be shown in meters, accurate to the nearest millimeter. Stations will be in 100-meter intervals (e.g., STA 3+75.289). Note the continued use of the plus (+) sign; since this is common practice, it will quickly indicate the number is a station dimension. For all stations within 100 meters of the reference point, the zero plus (0+) station designation will continue to be used. An example of how such a station will be indicated is STA 0+69.134.

The symbols for elevation and station are to be written in capital letters with no periods. Do not include the unit symbol after station and elevation designations; the unit of measure will be shown in a note.

Transition curves.—Figure 4-17 is an example of the curve radius. The curve radius is that distance from a vertex from which an arc can be drawn so that the arc will be tangent with each of two lines at their termination points. The circular arc which connects the two straight lines is called the transition curve.

The use of curve radius is the preferred design method for defining the transition curve for joining two straight line segments. The use of degree of curve is discouraged because an acceptable definition using metric units is not known, and it was not considered wise to unilaterally establish such a definition. The alternative of soft converting either the highway or railroad definitions was considered unacceptable. The radius is given in meters.

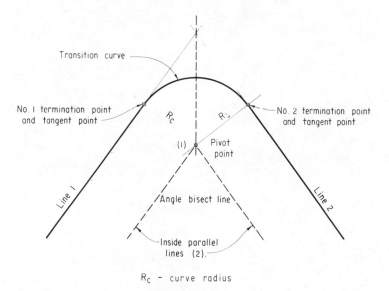

(1) The pivot point can be found by extending a perpendicular line from either termination point and intersecting with the angle bisect line or a perpendicular line from the other termination point.

(2) The pivot point and termination points can be established for a known curve radius (R_c) by running inside parallel lines a distance of R_c from each main line segment and determining the intersection point; this is the pivot point.

Having determined the pivot point, the termination points are then located by finding where the transition arc is tangent to the straight line segments.

Figure 4-17.—Plotting the transition curve. 40-D-6354

CHAPTER 4—USBR ENGINEERING APPLICATIONS   151

It should be noted that all the Bureau computer programs for survey and mapping have been metricated to accept the curve radius as input data.

Scale ratios.—All maps or site plans produced by or for the Bureau should be drawn to one of the scales listed in table 4-18.

Table 4-18.—*Map and site plan scale ratios*

1:100	1:1 000	1:10 000	1:100 000	1:1 000 000
1:200	1:2 000	1:20 000	1:200 000	1:2 000 000
1:500	1:5 000	1:50 000	1:500 000	1:5 000 000

Scales having ratios of 1:25 000 and 1:24 000 will be required to be used for scaling U.S. Geological Survey maps or when maps and site plans are produced to interface with other drawings having these scale ratios. Otherwise, the scale ratios listed in table 4-18 should be used.

The following are contour intervals (in meters) to be used on maps or site plans:

0.1, 0.2, 0.5, 1.0, 2, 5, 10, 20, 50, and 100

For information regarding the USGS metric mapping policy, see the additional information contained in the appendix.

International Size Paper

The future replacement of presently used paper sizes with international paper sizes[6] has been considered by the E&R Center Metric Committee. The sizes of paper being considered for replacement range from the 8- by 10-1/2-inch bond typing sheet to the 21- by 36-inch standard drawing sheet. Although there are no immediate plans to discard the presently used sizes, investigatory steps have been initiated to determine what international sizes are suitable and what problems or limitations will result from a particular size.

International paper sizes[7] consist of three series of paper sizes; these are the A-series, the B-series, and the C-series. Each series of paper sizes starts with one basic size; succeeding sizes are determined by a geometric relation with the previous size. Each paper series is described in the following:

A-size.—The basic paper size (A0) of the A-series has an area of 1 m^2; the ratio of length (L) to width (W) is the square root of 2.

The series progresses in size by halving the larger dimension, the length, until some arbitrary termination size is reached. Through this halving process, the $\sqrt{2}$ ratio of length to width is maintained. Table 4-19 lists the basic A-series sizes of paper along with the inch equivalents.

---

[6] International paper sizes are not an integral part of SI.
[7] ISO has adopted the A-series as the preferred series. The B-series and a modified C-series are also part of ISO R216. The ISO recommendation contains sizes not presented here; they are not universally accepted.

Table 4-19.—*A-series paper sizes*

Size designation	Metric[1] dimensions, mm			Customary equivalents, inches		
	W	x	L	W	x	L
A0	841	x	1189	33.11	x	46.81
A1	594	x	841	23.39	x	33.11
A2	420	x	594	16.54	x	23.39
A3	297	x	420	11.69	x	16.54
A4	210	x	297	8.27	x	11.69
A5	148	x	210	5.83	x	8.27
A6	105	x	148	4.13	x	5.83
A7	74	x	105	2.91	x	4.13
A8	52	x	74	2.05	x	2.91
A9	37	x	52	1.46	x	2.05
A10	26	x	37	1.02	x	1.46

[1] In the halving process, odd numbers divided by two are rounded down to the nearest whole millimeter.

B-size.—The basic size (B0) of the B-series has an area of $\sqrt{2}$ m^2. The ratio of length to width is the same as the A-series, $\sqrt{2}$. Table 4-20 lists the dimensions of the B-sizes of paper.

Table 4-20.—*B-series paper sizes*

Size designation	Metric dimensions, mm			Customary equivalents, inches		
	W	x	L	W	x	L
B0	1000	x	1414	39.37	x	55.67
B1	707	x	1000	27.83	x	39.37
B2	500	x	707	19.68	x	27.83
B3	353	x	500	13.90	x	19.68
B4	250	x	353	9.84	x	13.90
B5	176	x	250	6.93	x	9.84
B6	125	x	176	4.92	x	6.93
B7	88	x	125	3.46	x	4.92
B8	62	x	88	2.44	x	3.46
B9	44	x	62	1.73	x	2.44
B10	31	x	44	1.22	x	1.73

C-size.—The basic paper size (C0) of the C-series has an area of $\sqrt[4]{2}$ m^2, which equals 1.189 m^2. The ratio of length to width is $\sqrt{2}$. Table 4-21 lists the dimensions of the C-size of paper.

CHAPTER 4—USBR ENGINEERING APPLICATIONS

Table 4-21.—*C-series paper sizes*

Size designation	Metric dimensions, mm			Customary equivalents, inches		
	W	x	L	W	x	L
C0	917	x	1297	36.10	x	51.07
C1	648	x	917	25.53	x	36.10
C2	458	x	648	18.03	x	25.53
C3	324	x	458	12.72	x	18.03
C4	229	x	324	9.02	x	12.72
C5	162	x	229	6.36	x	9.02
C6	114	x	162	4.49	x	6.36
C7	81	x	114	3.18	x	4.49
C8	57	x	81	2.24	x	3.18
C9	40	x	57	1.57	x	2.24
C10	28	x	40	1.10	x	1.57

The paper from which the trimmed A-series paper sizes are cut is the RA and SRA mill sizes. The width dimensions of these sizes are listed in table 4-22.

Table 4-22.—*Mill paper widths*

Series designation	Widths, mm[1]			
RA	430	610	860	1220
SRA	450	640	900	1280

[1] Each series also has three sheet-size designations; base length is 860 and 900 mm for the RA and SRA series, respectively.

In selecting a paper size series, some of the problem areas under review are identified:

    a. *Storage*.—Are the present letter and drawing files properly sized to permit use of international size paper? The A-series typing size sheet (A4) will fit in most USBR file cabinets. The proposed drawing size sheets are A1 and B1; A1 paper fits in all drawing files and B1 fits in some drawing files. Most of the currently used notebooks will be too small for the ISO paper sizes.

    b. *Reproduction*.—Electrostatic copying machines used in Europe are capable of using the A4 size paper. Changeover cost for the present copying machines is unknown. The large photocopy and blueline machines will accept the A1 and B1 size drawing sheets. Some paper loss will occur because oversized paper will be required; this will have to be trimmed to match the original paper.

    c. *Nominal dimensions*.—In replacing the standard 21- by 36-inch drawing sheet, some questions arise as to the actual dimensions of the A1 and B1 sheets. The present standard drawing sheet has three distinct measurements for length and width; for each, there is the distance between trimmed edges, the distance between the trim lines, and the distance between the dark border lines. The present 21- by 36-inch dimensions are the lengths between the dark borderlines. A decision must be made as to whether the A1 sheet dimensions of 841 by 594 mm will apply to the outside edges, thus reducing the drawing area, or whether an oversized A1 sheet will be required such that these dimensions may apply to the borderlines.

Reading Metric Gages

Micrometers

Figure 4-18 illustrates a metric micrometer which will replace the present inch micrometer. The micrometer reading scales are the upper and lower sleeve markings and the thimble graduations. The metric micrometer spindle has 50 threads per 25 millimeters; one complete revolution of the thimble represents a spindle movement of 0.5 mm.

In reading the metric micrometer, the upper sleeve marks indicate 1 mm; the lower sleeve marks divide the upper sleeve marks and so represent 0.5 mm. The thimble is graduated into 50 equal markings with each graduation equaling 0.01 mm.

Referring to figure 4-18, note that the thimble edge is to the right of the 5-mm mark on the upper sleeve and is also to the right of the 0.5 mm on the lower sleeve. The "30" line on the thimble alines with the sleeve centerline. Therefore, the complete reading given in figure 4-18 is (5.0 + 0.5 + 0.30 =) 5.80 mm.

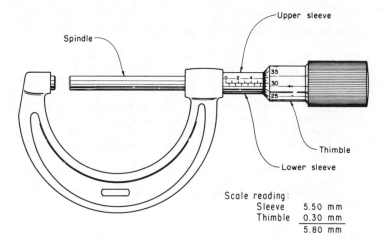

Figure 4-18.—Typical metric micrometer—graduated in hundredths of a millimeter.
40-D-6355

CHAPTER 4—USBR ENGINEERING APPLICATIONS

The previous example applies to a micrometer graduated in hundredths of a millimeter. A micrometer graduated in thousandths is read the same way except that a thousandths reading obtained from a vernier scale is added on. Figure 4-19 gives an exploded view of micrometer scale graduated in thousandths. Note that the 1-mm and 0.5-mm marks are both on the lower sleeve. The hundredths reading is still obtained from the thimble scale; the value read is the next lowest number to the sleeve centerline. The thousandths reading is obtained by determining which horizontal sleeve mark (vernier scale) most closely alines with a thimble graduation; in this case, it appears that the "8" line alines with the "23" thimble mark. The micrometer reading is (2.5 + 0.15 + 0.008 =) 2.658 mm.

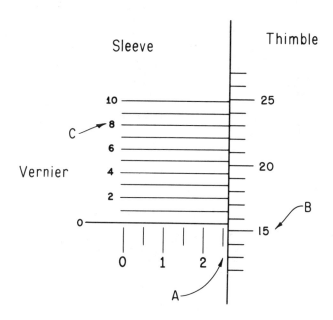

(A) Reading on the sleeve . . . . 2.5 mm
(B) Reading on the thimble . . . . 0.15 mm
(C) Reading on the vernier . . . . 0.008 mm

Figure 4-19.—A vernier scale for a micrometer graduated in thousandths of a millimeter. 40-D-6356

Vernier Scale

Vernier calipers are calibrated to read in units as small as 0.05 or 0.02 mm. The large divisions on the main scale will be in centimeters which are subdivided into 1-mm intervals, and the numbers marked on the vernier scale represent 0.10 mm. The precision at which the caliper will measure depends on the number of graduations on the vernier; 20 divisions are required for precision to the nearest 0.05 mm, and 50 divisions are required for precision to the nearest 0.02 mm. It is also possible that the main scale of the caliper will show finer graduations. A vernier scale with 25 divisions will achieve a precision of 0.02 mm if the main scale divisions represent 0.5 mm.

Figure 4-20 gives several examples of reading metric calipers.

Metric Dial Calipers

The metric dial caliper consists of two reading scales; refer to figure 4-21 for identification of the (1) main beam scale and (2) the dial face. In this example, the main beam scale is divided into 5-mm graduations; the large graduations represent 1 cm (10 mm). One complete revolution of the dial indicator represents 5 mm. The dial face will have five major divisions, each representing 1 mm; these major divisions are further graduated into 20 divisions for 0.05-mm readings, or 50 divisions for 0.02-mm readings.

Figure 4-21 illustrates a dial face caliper reading 36.25 mm.

Soil Classification

Sieve Sizes

The Bureau uses the ASTM E-11 sieve designation in prescribing the soil characteristics or gradation to be used in construction projects. Past practice has prescribed the use of screen numbers (openings per inch) or nominal inch-size openings in describing the sieve sizes which have always been metrically dimensioned. Table 4-23 shows a modified version of the ASTM E-11 sieve size table. The previously listed specifications paragraphs were revised to reflect the metric sieve sizes as listed by ASTM E-11.

One of the several ISO standards covering sieve sizes is ISO 565. This standard identifies the two principal series of sieve sizes, the R20 and R40, and lists the size designations associated with each. The ASTM E-11 standard includes almost all sizes from each series.

In SI, it is the recommended practice to use only one unit for a particular measure. This would indicate that all sieve sizes should be given in millimeters or micrometers, but not both.

However, ISO 565 and ASTM E-11 both list all sizes below 1 mm in micrometers and all sizes from 1 mm in millimeters (table 4-23). As this is the standard practice, it will be the method used by the Bureau; ASTM E-11 will be the principal standard.

CHAPTER 4—USBR ENGINEERING APPLICATIONS

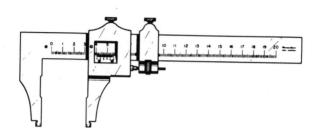

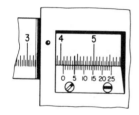

Main beam scale:
  Large numbers = 1 cm (10 mm)
  2nd level marks = 1 mm
  Halving marks = 0.5 mm
Vernier scale:
  25 divisions which divide
  each 0.5 mm into 0.02-mm
  graduations

Scale Reading:
  Main beam   40.50 mm
  Vernier   0.10 mm
               40.60 mm

(No. 5 mark on vernier
alines with 44-mm mark.
5 x 0.02 = 0.10 mm.)

Sheet 1 of 2

Figure 4-20.—Metric calipers with vernier scales. 40-D-6357-1

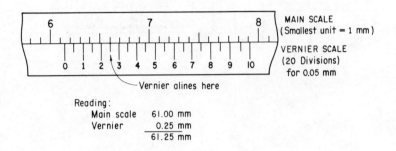

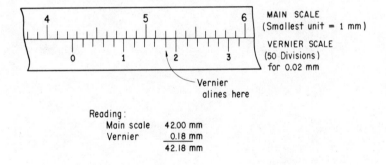

Figure 4-20.—Metric calipers with vernier scales. 40-D-6357-2

CHAPTER 4—USBR ENGINEERING APPLICATIONS

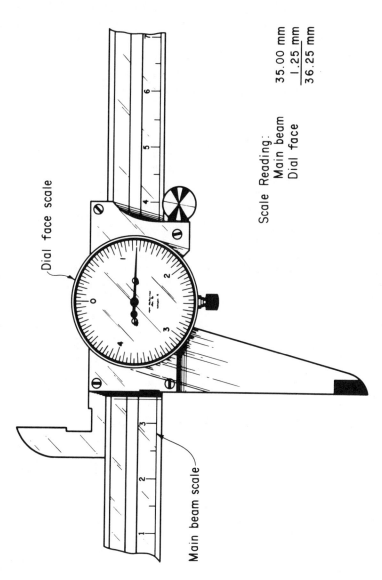

Figure 4-21.—Dial face metric caliper. 40-D-6358

Table 4-23.—*U.S. and ISO standard test sieves and soil gradations*

Customary	ASTM Standard Sieve Designations (Classification)	Metric	Permissible variance, plus or minus
5 in	Cobble	125 mm	3.7 mm
4.24 in	Cobble	106 mm	3.2 mm
4 in	Cobble	100 mm	3.0 mm
3-1/2 in	Cobble	90 mm	2.7 mm
3 in	Coarse gravel	75 mm	2.2 mm
2-1/2 in	Coarse gravel	63 mm	1.9 mm
2.12 in	Coarse gravel	53 mm	1.6 mm
2 in	Coarse gravel	50 mm	1.5 mm
1-3/4 in	Coarse gravel	45 mm	1.4 mm
1-1/2 in	Coarse gravel	[1]38.1 mm	1.1 mm
1-1/4 in	Coarse gravel	31.5 mm	1.0 mm
1.06 in	Coarse gravel	26.5 mm	0.8 mm
1 in	Coarse gravel	25.0 mm	0.8 mm
7/8 in	Coarse gravel	22.4 mm	0.7 mm
3/4 in	Coarse gravel	19.0 mm	0.6 mm
5/8 in	Fine gravel	16.0 mm	0.5 mm
0.530 in	Fine gravel	13.2 mm	0.41 mm
1/2 in	Fine gravel	12.5 mm	0.39 mm
7/16 in	Fine gravel	11.2 mm	0.35 mm
3/8 in	Fine gravel	9.5 mm	0.30 mm
5/16 in	Fine gravel	8.0 mm	0.25 mm
0.265 in	Fine gravel	6.7 mm	0.21 mm
1/4 in	Fine gravel	6.3 mm	0.20 mm
No. 3-1/2	Fine grav'	5.6 mm	0.18 mm
No. 4	Fine gravel	4.75 mm	0.15 mm
No. 5	Coarse sand	4.00 mm	0.13 mm
No. 6	Coarse sand	3.35 mm	0.11 mm
No. 7	Coarse sand	2.80 mm	0.095 mm
No. 8	Coarse sand	2.36 mm	0.088 mm
No. 10	Coarse sand	2.00 mm	0.070 mm
No. 12	Medium sand	1.70 mm	0.060 mm
No. 14	Medium sand	1.40 mm	0.050 mm
No. 16	Medium sand	1.18 mm	0.045 mm
No. 18	Medium sand	1.00 mm	0.040 mm

Table 4-23.—*U.S. and ISO standard test sieves and soil gradations*—Continued

ASTM Standard Sieve Designations			Permissible variance, plus or minus
Customary	(Classification)	Metric	
No. 20	Medium sand	850 μm	35 μm
No. 25	Medium sand	710 μm	30 μm
No. 30	Medium sand	600 μm	25 μm
No. 35	Medium sand	500 μm	20 μm
No. 40	Medium sand	425 μm	19 μm
No. 45	Fine sand	355 μm	16 μm
No. 50	Fine sand	300 μm	14 μm
No. 60	Fine sand	250 μm	12 μm
No. 70	Fine sand	212 μm	10 μm
No. 80	Fine sand	180 μm	9 μm
No. 100	Fine sand	150 μm	8 μm
No. 120	Fine sand	125 μm	7 μm
No. 140	Fine sand	106 μm	6 μm
No. 170	Fine sand	90 μm	5 μm
No. 200	Fine sand	75 μm	5 μm
No. 230		63 μm	4 μm
No. 270		53 μm	4 μm
No. 325		[3] 45 μm	3 μm
No. 400		38 μm	3 μm
No. 560	See footnote 2.	25 μm	—
No. 625		20 μm	—
No. 1250		10 μm	—
No. 1750		8 μm	—
No. 2500		5 μm	—
No. 5000		2.5 μm	—
No. 12000		1 μm	—

[1] ISO 565 lists this value as 37.5 mm; editorially changed by ASTM.
[2] Considered to be either silt or clay, depending on behavior characteristics.
[3] All values below 45 μm are considered supplementary sizes. Not all supplementary sizes are listed.

7m-1336 (4-77)
Bureau of Reclamation

LOG OF TEST PIT OR AUGER HOLE
FOR BORROW AND FOUNDATION INVESTIGATIONS

Feature ____ Project ____ Area Designation ____
Hole No. ____ Coordinates N ____ E ____ Ground Elevation ____ Approx. Dimensions1/ ____
Depth to Water Level1/ & 2/ ____ Method of Excavation ____ Date ____ Logged by ____

CLASSIFICATION SYMBOL	DEPTH (m)1/	SIZE AND TYPE OF SAMPLE TAKEN3/	CLASSIFICATION AND DESCRIPTION OF MATERIAL (SEE DESIGNATION E-3, EARTH MANUAL, FOR METHOD OF DESCRIPTION; GIVE GEOLOGIC AND IN-PLACE DESCRIPTION FOR FOUNDATION INVESTIGATIONS)	PERCENTAGE OF COBBLES AND BOULDERS4/				
				VOLUME OF HOLE SAMPLED (m^3)5/	MASS OF 75 TO 125 mm SAMPLED (kg)3/	% BY VOLUME OF 75 TO 125 mm6/	MASS OF PLUS 125 mm SAMPLED (kg)3/	% BY VOLUME OF PLUS 125 mm6/

REMARKS:

NOTES:
1/Report to nearest 0.01 m
2/Record after water has reached its natural level and give date of reading
3/Report to nearest 0.5 kg
4/Applicable only to borrow pits and to foundations which are potential sources of construction materials
5/Report to nearest 0.001 m^3
6/ (1) $\% = \frac{\text{(mass of rock sampled)}}{\text{(Bulk density) X (cubic metres of hole sampled)}} \times 100$
(2) Record bulk density in Remarks, stating how obtained (measured or estimated)
Report to nearest 1%

GPO 837-864

Figure 4-22.—Test pit log sheet. 40-D-6359

CHAPTER 4—USBR ENGINEERING APPLICATIONS

Particle Classification

Table 4-24 shows the type of soil classification to be used for a test pit or auger hole field investigation. Since this is a visual classification, it would be very inappropriate to use an increment smaller than 5 millimeters. Figure 4-22 is the data form used for this type of field investigation.

Table 4-24.—*Visual classification—description of particle size*

Range of maximum particle sizes	Round to nearest	Example
0 to 5 mm	Describe as a fine, medium, or coarse sand	
5 to 50 mm	5 mm	5, 10, 15, etc.
50 to 300 mm	25 mm	100, 150, etc.
300 to 1000 mm	100 mm	500, 600, etc.

Use of the SI (International System of Units) in Data Processing

Introduction

The use of the SI (International System of Units) is of interest to data processing personnel because of the increased emphasis on metrication and the limitations that most data processing equipment place on presentation of metric units.

In most cases, the available data processing character set is limited, thus special rules must be applied to the representation of metric units and prefixes in computer input and output.

This summary of the various rules concerning the representation of metric units is intended as a guide to be tempered with good judgment in presenting the SI system of units to computer users.

Character Set

The characters 0 through 9, A through Z (capital letters only), hyphen and minus sign (-), plus (+), slash (/), period (.), and "space" are used without the capability to use digits and signs in superscript position. Greek letters such as Ω for electric resistance and μ for micro are not available.

Rules to Represent SI Units in Print Output

The following six rules are presented to indicate the uniform method to be used in presenting SI units of measure in ADP input/output:

Rule 1.—Use the space character to separate the numeric value and the unit symbol or name.

Examples:

10 M ("ten-space-em") to designate 10 meters.
2 M2 ("two-space-em-two") to designate 2 square meters.

Rule 2.—To indicate multiplication of units, a period (.) between the unit symbols (combined with prefix or not) is necessary.

Examples:

PA.S to designate pascal second, the unit of dynamic viscosity.
N.M to designate newton meter. The period between the two letters is intended to avoid confusion which could occur between N.M (newton meter) and NM (nanometer).

Rule 3.—To indicate division of units, the numerator and the denominator are separated by a slash (/); alternatively, the denominator may be expressed with a negative exponent (see rule 5).

Examples:

M/S ("em-slash-es") or M.S-1 ("em-period-es-hyphen-one") for meter per second.

Rule 4.—Positive exponents are indicated by the respective numerals without any sign, directly after the representation of the unit.

Example:

M2 ("em-two") for square meter and meter squared.

Rule 5.—Negative exponents are indicated by a minus sign followed by the respective numeral, both together directly after the representation of the unit.

Example:

M-3 ("em-hyphen-three") for one divided by M to the third power, or one per cubic meter ($1/m^3$).

Rule 6.—A prefix representation is combined with a unit representation to form a new unit representation which can be raised to a power with positive or negative exponents and which can be combined with other unit representations to form representations or compound units. There is no separator or space between the prefix representation and the unit representation.

CHAPTER 4—USBR ENGINEERING APPLICATIONS

Examples:

CM2 ("see-em-two") for square centimeters. KN/M2 ("ka-en-slash-em-two") or KN.M-2 ("ka-en-period-em-hyphen-two") for kilonewtons per square meter.

Notes

1. A prefix is not allowed to stand alone, without combining with a unit. For example, T alone means tesla (magnetic flux density, magnetic induction) but not tera (ten to the twelfth power). Compound prefixes should not be used; for example, NM (nanometer) should be used instead of MUM (millimicrometer).

2. In the case of the SI base unit kilogram, which contains a prefix in its name, the representations for decimal multiples and submultiples are formed by use of the corresponding prefix representation together with the representation of the unit gram; for example, the representation for one millionth of a kilogram (one thousandth of a gram) is denoted by MG (milligram), not UKG (microkilogram).

Representation of Units

The following four tables list the ADP symbol representation for SI units. These symbols, when used per the preceding six style rules, provide sufficient flexibility for ADP equipment to write the SI language.

Table 4-25.—*SI base units*

Name	ADP representation
ampere	A
candela	CD
kelvin	K
kilogram	KG
meter	M
mole	MOL
second	S

Table 4-26.—*SI supplementary units*

Name	ADP representation
radian	RAD
steradian	SR

Table 4-27.—*Derived SI units with special names*

Name	ADP representation
coulomb	C
farad	F
henry	H
hertz	HZ
joule	J
lumen	LM
lux	LX
newton	N
ohm	OHM
pascal	PA
siemens	SIE
tesla	T
volt	V
watt	W
weber	WB

Table 4-28.—*Other units*

Name	ADP representation
degree (angle)	DEG
degree Celsius	CEL
gram	G
liter	L
minute (angle)	MNT
second (angle)	SEC

Prefixes

Table 4-29 lists the ADP symbols representing the SI prefixes. The use of these prefixes is covered by rule 6.

Table 4-29.—*Representation of prefixes*

Prefix	Multiplier*	ADP representation
exa	18	E
peta	15	P
tera	12	T
giga	9	G
mega	6	MA
kilo	3	K
hecto	2	H
deca	1	DA
deci	-1	D
centi	-2	C
milli	-3	M
micro	-6	U
nano	-9	N
pico	-12	P
femto	-15	F
atto	-18	A

* The multiplier represents the power to which ten is raised. Examples: "12" means ten raised to the twelfth power (10^{12}) and "-3" means ten raised to the minus 3 power (10^{-3} or one thousandth).

Decimal Marker

A period is used as the decimal marker. In written material, a zero must precede the decimal marker. The Bureau's ADP computer language presently is not able to print output in this format uniformly; the extensive modifications to the programs and/or the compilers needed to accomplish this is impractical.

Example: Seventy-five hundredths will be printed .75.

Grouping of Digits

Numerical values for ADP equipment (input and output) will be blocked; that is, there will be no spaces or commas between the digits, only the decimal marker will appear, as appropriate. The rule which requires that large numbers be written in three-digit groups presently will not apply to ADP output. At the time this feature becomes a standard part of the compilers, it will be applied in new applications wherever practicable.

Chapter V

UNIT AND FORMULAE CONVERSIONS

Introduction

The information provided in this chapter is to aid in the conversion to SI units. This information is required during the "soft" conversion transition period; once complete "hard" conversion has been accomplished, these data will be of limited usefulness. A basic point of philosophy is that actually using SI units is the only way the system can be effective; working with customary units and then converting is not an acceptable alternative. The latter situation will preclude realization of the benefits available from using SI.

Dimensions and Dimensional Analysis

Dimensional representation is a method of expressing units of measure with dimensional symbols which delineate basic physical quantities. There are several systems or sets of dimensional symbols which have been developed; the system used in this manual is the $LTMQ\theta$ system. See table 5-1 for the symbol designations. Table 5-2 lists several physical quantities and the dimensions which represent them. In the conversion tables, the dimensional symbols are also shown for many of the physical quantities.

Table 5-1.—*Basic dimensional units*

length	L	electric current	Q[1]
time	T	plane angle	θ
mass	M		

[1] In most dimensional analysis systems, Q usually represents electric charge (coulombs). To comply with the SI designation of electric current (amperes) as a base unit, table 5-2 listings of quantities involving Q show the dimensional representations with electric charge = Q x time. This was accomplished by substituting QT wherever Q was previously used.

Theoretically, a dimensional symbol is needed for each of the seven base units and the two supplementary units,[1] but practice has shown that these five symbols will be adequate for the most engineering problems encountered by the Bureau. The importance of these dimensional symbols and dimensional analysis is that they are very useful in determining if a formula/equation needs modification before it is used with SI units, and if so, what changes are necessary. Most changes which may be required in a formula/equation will occur with multiplication coefficients and constants.

---

[1] These nine units were established as being dimensionally independent. This means none of them can be defined in terms of the others.

Table 5-2.—*Compound dimensional units*

Quantity	Dimensions	Quantity	Dimensions
acceleration	L/T^2	capacitance	$T^4 Q^2/ML^2$
area	L^2	charge	QT
force	ML/T^2	electric displacement	QT/M^2
momentum	ML/T	inductance	$ML^2/Q^2 T^2$
work	ML^2/T^2	magnetic flux	ML^2/QT^2
		permeability (elect)	ML/QT
entropy	ML^2/T^2	resistivity	$ML^3/Q^2 T^3$

There are two basic rules for changing a formula/equation:

1. If a constant is dimensionless and the variables of the formula/equation are all in "basic units," the formula can be used without change. Basic units refer to feet, pounds, and/or seconds in the U.S. customary system, and to meters, kilograms, and/or seconds in SI.

2. Changes will be required if a constant has dimensions, or if some of the variables or the resultant are not in basic units (e.g., acre-feet or gallons instead of cubic feet, or liters instead of cubic meters).

The following examples illustrate the use of dimensional analysis in changing a formula from U.S. customary units to SI units:

Example 1: Rectangular weir

$$Q = 3.33 (L - 0.2h)(h)^{1.5}$$

where

Q = flow in cubic feet per second (ft^3/s)
L = crest length in feet (ft)
h = head in feet (ft)
3.33 = coefficient of discharge, C

Dimensionally, the formula is represented:

$$L^3/T = C \times L \times L^{1.5}$$

$$L^3/T = C \times L^{2.5}$$

$$\therefore C = L^{0.5}/T$$

To achieve dimensional balance, the discharge coefficient must have the dimensions shown. Before SI units can be substituted, the discharge coefficient must be changed; this can be done in the following manner:

CHAPTER 5—UNIT AND FORMULAE CONVERSIONS

Establishing the following proportions:

$$\frac{C_{SI}}{C} = \frac{m^{0.5}/s}{ft^{0.5}/s}$$

After dividing out s, substituting 3.33 for C, and transposing, the result is:

$$C_{SI} = 3.33\sqrt{m/ft}$$

To convert feet to meters, multiply by 0.3048. (Caution, this is not the same as saying 1 m = 0.3048 ft, which is incorrect.)[2] Substituting (m = 0.3048 x ft) into the former equation:

$$C_{SI} = 3.33\sqrt{\frac{0.3048 \times ft}{ft}}$$

$$= 3.33\sqrt{0.3048}$$

$$= 1.838$$

Rewriting the formula for SI units, it is now:

$$Q = 1.838\,(L - 0.2h)(h)^{1.5}$$

where

Q = flow in cubic meters per second (m^3/s)
L = crest length in meters (m)
h = head in meters (m)
1.838 = coefficient of discharge, $\sqrt{m/s^2}$

An alternative method which will give the same result follows:

$ft^3/s = 3.33 \times ft \times ft^{1.5}$ [U.S. customary]

$m^3/s = C_{SI} \times m \times m^{1.5}$ [SI metric]

Dividing the first equation by the second, results in:

$$\frac{ft^3/s}{m^3/s} = \frac{3.33}{C_{SI}} \times \left(\frac{ft}{m}\right)^{2.5}$$

---

[2] The feet to meters conversion equation, m = 0.3048 x ft, represents the conversion of the numerical value represented by "ft" to the numerical value represented by "m." The multiplication factor has the dimensions of "m/ft;" therefore, if the multiplication were completed for 1 ft, the result would be 1 ft = 0.3048 m.

$m = 0.3048 \times ft$ [feet to meters conversion]

$m^3/s = ft^3/s \times 0.02832$ [cubic feet to cubic meters conversion]

Substituting:

$$\frac{ft^3/s}{ft^3/s \times 0.02832} = \frac{3.33}{C_{SI}} \left(\frac{ft}{0.3048 \times ft}\right)^{2.5}$$

$C_{SI} = 3.33 \, (19.4968) \, (0.02832)$

$\quad\quad = 1.838$

Example 2: Orifice discharge

$$Q = C_d A \sqrt{2gh}$$

where

Q = flow in cubic feet per second (ft^3/s)
A = orifice area, square feet (ft^2)
g = gravitational acceleration, feet per second squared (ft/s^2)
h = head in feet (ft)
C_d = coefficient of discharge

Dimensionally:

$$L^3/T = C_d \times L^2 \times \sqrt{L/T^2 \times L}$$

$$= C_d \times L^2 \times L/T$$

$$L^3/T = C_d \times L^3/T$$

Therefore, C_d is dimensionless and the formula can be used with SI units as it stands.

Q = flow, cubic meters per second (m^3/s)
A = orifice area, square meters (m^2)
g = gravitational acceleration, meters per second squared (m/s^2)
h = head, meters (m)
C_d = discharge coefficient, dimensionless

Example 3: Pipe constriction head loss

$$H = 0.375 \frac{V^2}{2g}$$

CHAPTER 5—UNIT AND FORMULAE CONVERSIONS

where:

H = head loss at a sudden pipe constriction in feet of water (meters of water)
V = velocity in feet per second (meters per second)
g = gravitational acceleration, ft/s^2 (m/s^2)
0.375 = coefficient constant

Dimensionally:

$$L = 0.375 \times (L/T)^2 / (L/T^2)$$

$$L = 0.375 \times L$$

The coefficient is dimensionless. The SI units shown in parentheses may be used without changing the formula.

Example 4: Chezy-Manning formula

$$V = \frac{1.486}{n} r^{0.667} S^{0.5} \quad \text{[U.S. customary]}$$

where:

V = velocity in feet per second (meters per second)
r = hydraulic radius in feet (meters)
S = hydraulic gradient, dimensionless
n = roughness coefficient

Dimensionally:

$$L/T = \frac{1}{n} \times L^{0.667} \times 1$$

$$\sqrt[3]{L/T} = 1/n$$

$$n = T/\sqrt[3]{L}$$

Conversion will be required; establishing the SI/U.S. customary ratio:

$$\frac{n_{SI}}{n} = \frac{s/\sqrt[3]{m}}{s/\sqrt[3]{ft}}$$

$$n_{SI} = n\sqrt[3]{ft/m}$$

$$m = 0.3048 \times ft \quad \text{[feet to meters conversion]}$$

$$n_{SI} = n \sqrt[3]{\frac{ft}{0.3048 \times ft}}$$

$$= n\sqrt{3.28084}$$

$$= n(1.486)$$

The formula now becomes:

$$V = \frac{r^{0.667}\sqrt{S}}{n} \qquad \text{[SI metric]}$$

The n values which have always been used are still acceptable. They were developed for metric units and the 1.486 coefficient has been the modification factor permitting the use of the U.S. customary units.

In these four examples, only basic units were considered in the formula modifications. If it is desired to use other units, the principles and methods still apply and will permit successful modifications for whatever units are desired for each quantity. While the nonbasic units may be more convenient measures in SI and the U.S. customary system, it is generally recommended that base units be utilized in all formulae. The nonbasic measures should be converted to a base unit along with the appropriate multiplier and used in the formulae in that form. Several examples of formula modification which involve nonbasic units are included in chapter VI.

Guidelines for Conversion and Rounding of Numerical Values

Definitions

Before discussing particular considerations and procedures to be used in the conversion and rounding process, a few definitions are listed below (table 5-3) which are presented to assure a good understanding of the succeeding material.

Table 5-3.—*Definitions of terminology*

Accuracy	—The error-free level or degree of conformity to which a calculated value corresponds to a standard or specified value. The smaller the error, the greater the accuracy.
Precision	—The extent to which a measured value can be reproduced or repeated, or the degree of mutual agreement between individual measurements; i.e., very fine increments of measure. The variation implied on the statement of the numerical value.
Significant digit	—Any digit that is necessary to define a specific value or quantity for its intended accuracy.
Tolerance	—The allowable deviation from the specified dimension, the upper and lower limits between which a dimension must be maintained.
Approximate value	—A value that is not exact but very near the correct value.

CHAPTER 5—UNIT AND FORMULAE CONVERSIONS 175

Table 5-3.—*Definitions of terminology*—Continued

Nominal value	—A convenient numerical designation existing in name only. Examples of this include the 2 by 4 stud and 1-inch pipe; none of these dimensions can be found on the actual item.
Rationalized value	—A value representing a multiple of a selected measurement module, similar to nominal value.
Rounded value	—A value that is not exact but contains enough significant digits to "adequately" define the measured dimension.

Conversion and Significant Digits

A measurement quantity consists of a numerical value and a measurement unit; the measurement unit may be one from several measurement systems. In converting from one measuring system to another, establishing the procedure for determining how many decimal positions should be retained can be most difficult. The retention of too many decimal positions results in awkward values indicating extended precision; the dropping of too many significant digits results in a loss of needed accuracy. All conversions should be handled logically, giving careful consideration to the intended precision of the original quantity. The intended precision is usually established by specific tolerance or by some particular knowledge of the original quantity. The initial step in a conversion is to determine what precision is required to assure accuracy that is neither exaggerated nor sacrificed. As mentioned, determining what number of significant digits should be retained may be very difficult unless some prior knowledge exists regarding the original measurement procedure. To exemplify this, consider the following examples:

(1) There is a reported length of 75 ft. The exact metric conversion is 22.86 m.

(2) Atmospheric pressure has a nominal value of 14.7 lb/in^2. The exact metric conversion for the standard atmosphere is 101 325 Pa or 101.325 kPa.

In the first example (1), the reported length of 75 ft may be a value rounded to the nearest foot or a value rounded to the nearest 5 ft. In the first case, it would be appropriate to present the metric value to the nearest 0.1 m (i.e., 22.9 m); in the latter case, rounding to the nearest whole meter is appropriate (i.e., 23 m).

The second example (2) presents a situation where a nominal value is being described; this would indicate that fewer significant digits are required to express the metric equivalent. For example, 101 kPa might seem to be a good number; however, the prime consideration is the relatively limited range of atmospheric pressure, which makes the first decimal position important. Considering this, 101.3 kPa would make a better choice for the nominal SI value for atmospheric pressure.

A topic briefly discussed in table 5-3 is significant digits; any of the digits, 1 through 9, is a significant digit. The status of a zero is ambiguous because the zero may be used to indicate precision or it may be used to indicate the magnitude of a number. Examples of the various conditions in which the zero may be used are listed below:

	Integer	Decimal	
(a)	007	0.007	NOT SIGNIFICANT
(b)	2008	0.2008	SIGNIFICANT
(c)	150	0.150	?

The status of the right-hand zero in item (c) is ambiguous; additional information about the numerical value is required before the zero's status can be clarified. The following rules can be useful in writing numbers which involve the use of potentially ambiguous zeros:

1. Zeros at the end of integers can be dropped and replaced by a power of 10 if the zeros are not significant.

Example: $128\,000 \rightarrow 128 \times 10^3$

If one or two zeros are significant, use a corresponding number of decimal position zeros to indicate this.

Example: $128\,000 \rightarrow 128.0 \times 10^3$ (1 significant zero)
  $\rightarrow 128.00 \times 10^3$ (2 significant zeros)

Numbers such as 100 and 10 should not use scientific notation unless it is the format established for all numbers. For 10, 20, etc., consider the zero to be significant; for 100, 200, etc., consider only the zero in the tens position to be signficiant.

Note: Do not use terms such as billion, trillion, etc. The use of thousands and millions is acceptable. It is preferred that the SI prefixes be used with the appropriate unit names such that the use of scientific notation or numerical nouns is not necessary; this does not apply to pure numbers.

2. Zeros at the right-hand end of a decimal number must be considered significant unless specifically indicated otherwise. Zeros should not be added at the right-hand end of a decimal number unless they are significant.

Example: 0.654 000 (6 significant digits)

Manipulation of Numerical Data

Multiple levels of significant digits.—Various data may be acquired from diverse sources and each element may reflect differing degrees of refinement. Certain rules must be followed when such data are to be subjected to the four basic mathematical operations (addition, subtraction, multiplication, and division).

For addition and subtraction, the rules are:

CHAPTER 5—UNIT AND FORMULAE CONVERSIONS   177

1. The answer shall contain no more significant digits (or decimal positions) than contained in the **least** precise number.

Examples: (a)
```
      1.267 6
      3.21      ← Least precise number, two decimal
      5.361         positions
      1.360 10
     11.198 7    Round to 11.20
```

(b)
```
     +5.7438
     -3.1       ← Least precise number, one decimal
     +6.127        position
     -2.11
      6.6608    Round to 6.7
```

2. Before performing the mathematical operation, it is recommended that individual numbers, other than the least precise number, should be rounded to have one more significant digit than the least precise number. Applying this rule to the previously used examples, they now become:

(a)
```
      1.268
      3.21
      5.361
      1.360
     11.199     Which rounds to 11.20
```

(b)
```
     +5.74
     -3.1
     +6.13
     -2.11
      6.66      Which rounds to 6.7
```

Following this rule produces the same results but eliminates the necessity of carrying numbers with a large number of decimal positions which are eventually dropped in the final rounding process.

For multiplication and division, the rules are:
1. The product or quotient should not contain any more significant digits than are contained in the number with the fewest significant digits.
2. Do not use preliminary rounding.

Examples: (a) $6.8714 \times 2.6 = 17.865\ 64$ Round to 17.9
(b) $6.871 \times 2.60 = 17.8646$ Round to 17.86
(c) $6.2 \div 3.457 = 1.793\ 462^+$ Round to 1.8
(d) $6.208 \div 3.46 = 1.794\ 219^+$ Round to 1.79

Infinite number of significant digits.—Numbers representing exact counts or defined fractions are treated as if they consist of an infinite number of significant digits. When exact values are subjected to basic mathematical operations, no rounding is involved.

When these values are involved in computations with measured or estimated values, the resultant should contain the same number of significant digits as the measured or estimated value. The number of significant digits in values which are measured or estimated determines the level of implied precision in the computed answers; the answer can be no more precise than the least precise element.

Rounding of Data

When a numerical value is to be rounded to fewer significant digits than the total number available, the following procedure should be followed:

When the first digit dropped is:	The last digit retained is:	Examples
< 5	Unchanged	2.44 →2.4
> 5	Increased by 1	2.46 →2.5
5 followed by zeros	Unchanged if even, increased by 1 if odd	2.450 →2.4 2.550 →2.6

Rounding of Converted Values Without Tolerances

Nominal data are generally not critical and discretion may be used in rounding the converted values. However, data which represent measurements of estimated values often require greater consideration regarding the implied or intended precision of the value.

The most accurate equivalents in conversions are obtained by multiplying the numerical value by the conversion factor exactly as listed and then rounding the product. Preliminary rounding should be avoided as necessary precision may be sacrificed.

When a tolerance is not given and no knowledge of the original measure is available, the intended precision may be assumed to be plus or minus one-half unit of the least significant digit in which the value is given. Rounding of the original value is being assumed; one-half unit of the least significant digit represents the error limit of the original process. An example of this is 6.95; one-half of the last significant decimal position is $\left(\frac{0.01}{2}=\right)$ 0.005. (This establishes the limits on this number as 6.945 and 6.955.) Assuming the 6.95 is a measured value in inches and this value is being converted to millimeters: The 6.95 inch equals 176.53 mm; the intended precision of 0.005 inch converts to 0.127 mm. Therefore, the original value can be rounded to the nearest tenth of a millimeter; the result is 176.5 mm.

Two circumstances arise which may invalidate the use of this procedure. First, significant zeros may be omitted in the presentation of a value. The

CHAPTER 5—UNIT AND FORMULAE CONVERSIONS 179

dimension of 2.5 inches may mean about 2-1/2 inches, or it could be a shortened expression of an exact measure of 2.500 inches. The extra zeros in the latter expression are very significant in expressing the intended precision but do not affect the actual value. Secondly, quantities may be expressed in digits which are not intended to be significant. An example of this is the rough dimension, 7/16 inch; the exact decimal equivalent is 0.4375 inch. It is advisable that noncritical dimensions such as this be initially rounded (i.e., 0.44 inch) to avoid a greater implied precision than intended.

When the preceding procedure is more complicated than necessary, refer to table 5-4. When a maximum error difference is known, this table can be used to determine how the converted value should be rounded; the table is written for a maximum difference of 1 percent, but can be adjusted if necessary.

Rounding of Converted Values with Tolerances

The number of digits or decimals in the number expressing any quantity does not necessarily imply the accuracy of the measurement. Where it is required to indicate accuracy, limits of size or tolerances must always be given.

When accuracies or tolerances are expressed as percentages or proportional parts (e.g., 0.1 percent, 1 part in 10^5) rather than absolute values, the converted dimensions should be rounded and given to the same order of accuracy. Conversions based on this principle give accurate results simply. This also eliminates confusion in selecting the number of significant digits to be retained in the converted value and the magnitude of the deviation between the value and the original.

When dimensions are given with tolerances, the tolerances provide a good indication as to what is the intended precision. The tolerance is greater than the intended precision; for example, a tolerance of ±0.008 indicates a precision of 0.0005. Even when tolerances are supplied, the numerical values may still require careful examination and the use of judgment. To exemplify this, consider the following two dimensions:

(1) 1.055 ± 0.003 inch
(2) 3.625 ± 0.0625 inch (i.e., 3-5/8 ± 1/16 inch)

The first dimension is obviously quite precise and the implied precision indicated by the number of significant decimal positions is $\left(\frac{0.001}{2}=\right)$ 0.0005 inch. The second number represents a number measured to the nearest 1/8 inch with a tolerance of ±1/16 inch; this is certainly not a precision measurement, only a decimalized equivalent of a noncritical dimension.

There are several procedures which may be used to establish the implied precision or to determine the number of decimal positions in the metric equivalent. A general rule is to assume that the implied precision of a toleranced value is equal to 10 percent of the total tolerance. Table 5-5 expresses this guide differently, but the results are the same. This rule provides that the least significant digit retained in the conversion is no larger than one-tenth the converted tolerance.

Table 5-4.—*General rounding of data*[1]

Numerical range of data		Round to nearest[1,2]
Equal to or greater than	But less than	
0.000 5	0.002 5	0.000 01
0.002 5	0.005	0.000 05
0.005	0.025	0.000 1
0.025	0.05	0.000 5
0.05	0.25	0.001
0.25	0.5	0.005
0.5	2.5	0.01
2.5	5	0.05
5	25	0.1
25	50	0.5
50	250	1
250	500	5
500	2 500	10
2 500	5 000	50
5 000	25 000	100
25 000	50 000	500
50 000	250 000	1 000
250 000	500 000	5 000
500 000	2 500 000	10 000
2 500 000	5 000 000	50 000
5 000 000	25 000 000	100 000
25 000 000	50 000 000	500 000
50 000 000	250 000 000	1 000 000
250 000 000	500 000 000	5 000 000
500 000 000	2 500 000 000	10 000 000
2 500 000 000	5 000 000 000	50 000 000

[1] Use of this table will result in a maximum difference of 1 percent between the actual value and the rounded value.

[2] To obtain an error percentage other than 1 percent, multiply the right-hand column by a factor of n (n is the ratio of the desired error percentage to the 1 percent). Example: For a maximum error percentage of 0.3 percent, multiply by 0.3.

Table 5-5.—*Rounding converted tolerance in inches to millimeters*

Total tolerances in inches	Number of decimal positions for millimeters
$1 \times 10^{-5} < x < 1 \times 10^{-4}$	5
$1 \times 10^{-4} < x < 1 \times 10^{-3}$	4
$1 \times 10^{-3} < x < 1 \times 10^{-2}$	3
$1 \times 10^{-2} < x < 1 \times 10^{-1}$	2
$1 \times 10^{-1} < x < 1$	1

Table 5-6 was originally developed by Erisman [16] to aid in the interfacing problems between hardware designed in U.S. customary units and metric units; this table provides conversion guidelines for toleranced dimensions as well as limit dimensions. The principal feature of this table is that it considers the conversion discontinuities associated with certain values.

Table 5-6.—*Step procedure for determining decimal positions when converting precision unit dimensions and tolerances (U.S. customary to SI metric)*

Condition	Number of decimal positions in the conversion	U.S. customary (in)	SI metric (mm)
For ± tolerances other than 1×10^{-n}	One less than inch tolerance	0.646 ± 0.012 1.682 ± 0.14	[1] 16.41 ± 0.30 42.7 ± 3.6
For tolerances that equal 1×10^{-n} (e.g., 0.1, 0.01, 0.001)	Same as inch tolerance	0.742 ± 0.001 1.460 ± 0.01	[1] 18.847 ± 0.025 37.08 ± 0.25
When the difference between max./min. limit dimensions is other than 1×10^{-n} or 2×10^{-n}	One less than inch limit dimensions	$0.062/0.056$ $\Delta = 0.006$ $0.140/0.132$ $\Delta = 0.008$	[2] $1.57/1.43$ $3.55/3.36$
When the difference between max./min. limit dimensions equals 1×10^{-n} or 2×10^{-n}	Same as inch dimensions	$0.806/0.804$ $\Delta = 0.002$ $1.122/1.112$ $\Delta = 0.01$	[2] $20.472/20.422$ $28.498/28.245$

Courtesy of Machine Design

[1] Unequal tolerances are converted individually; an insignificant zero may be required after conversion to obtain an equal number of tolerance digits.

[2] For maximum/minimum limit values, the conversion for the maximum limit value should be rounded down (e.g., 2.8102 → 2.81 and 1.9678 → 1.967 or 1.96). The minimum limit conversion value should be rounded upward (e.g., 2.8102 → 2.811 or 2.82 and 1.9678 → 1.968 or 1.97).

Converting Temperatures

Temperatures expressed in a whole number of degrees should normally be converted to the nearest 0.5 °C or 0.5 K. The conversion of temperature tolerances should be in accordance with table 5-7.

Table 5-7.—*Conversion of temperature tolerances*

Tolerance, °F	Tolerance, °C or K
± 1	± 0.5
± 2	± 1.0
± 5	± 3.0
±10	± 5.5
±15	± 8.5
±20	±11.0
±25	±14.0

Just as with other quantities, the number of significant digits retained will depend upon the implied accuracy of the original value. An example of this is:

$$135 \pm 5 \,°F$$

The implied accuracy is estimated to be 2 °F. The 2 °F implied accuracy converts to a 1.111 °C implied accuracy for the Celsius value; this indicates the Celsius conversion should be to the nearest 1 °C. Therefore, the exact conversion (57.2222 ± 2.7778 °C) can be rounded to 57 ± 3 °C.

The Conversion Procedure

General.—The conversion procedure can be summarized by the following four steps:
1. Determine the implied precision:
   a. Plus and minus one-half unit of the least significant digit, or
   b. Ten percent of total tolerance, or
   c. Based upon knowledge of the original measurement (e.g., a value measured to the nearest 1/16 inch corresponds to a metric value to the nearest millimeter).
2. Convert the dimension, the implied precision, and the tolerance (if applicable) using the conversion factors listed in the appropriate table. It is important that neither the conversion factor nor the presented values be rounded before the multiplication is performed; accuracy will be sacrificed if the rounding process is started prematurely.[3]
3. Select the smallest number of decimal positions necessary to reflect the required precision as determined in step 1.

---

[3] Preliminary rounding of noncritical fractions expressed in thousandths is the exception to this rule.

CHAPTER 5—UNIT AND FORMULAE CONVERSIONS

4. As previously stated, round the converted numbers to the specified number of decimal positions according to the following rules:

When the first digit dropped is:	The least digit retained is:
<5	Unchanged
>5	Increased by 1
5 followed by zeros	Unchanged if even, increased by 1 if odd

Rounding of maximum/minimum limits.—The preceding rules are modified slightly when converting toleranced dimensions written as maximum and minimum limits. When judgment indicates that these values are absolute limit, a maximum quantity must be rounded downward and a minimum quantity is rounded upward. This method may also be used, if desired, for dimensions originally written as a mean value with a plus and minus tolerance; determine the upper and lower limits and proceed accordingly.

Miscellaneous rounding consideration.—Certain conditions or considerations may exist such that the previously listed guidelines for rounding of converted values may not satisfy the requirements. Such factors as safety and cost may alter circumstances; for example, thickness dimensions for steel plate should be rounded upward to assure adequate strength, but if rounded to too few significant digits, an unnecessary cost penalty would be incurred.

When feasible, the rounding of metric equivalents should be to rational, convenient numbers. Such conversions should be identified as approximate if the use of such a nominal value presents any possible difficulties.

Rationalized Metric Values

As the Bureau and all major industries evolve from the "soft" use of SI to the actual introduction of metrically designed and dimensioned materials and supplies, decisions regarding standard sizes will be required in all sectors of the economy. Ideally, the new metric sizes will be simple and easily remembered values. As an example, it is not practicable to dimension wood paneling with the soft conversion values of 1220 by 2440 mm, or, the exact value of 1219.4 by 2438 mm. Logic dictates that the wood products industry should adopt a 1200- by 2400-mm standard panel size. Similar changes will occur in all sectors of the economy; the method used to establish preferred sizes must be based on considerations relating to the quantity being reviewed and the degree of accuracy required.

There are several methods currently being used by industry in establishing product sizes or a series of sizes. The use of one particular metric module selection method cannot be recommended over another as each provides certain advantages depending on the product or industry involved.

The halving sequence.—The halving sequence is based on the $(1/2)^n$ multiplier. This multiplier is applied to a basic product size to establish successive product dimensions. The superscript, n, represents each successive level; the first reduction (n = 1) is 0.5 of the basic size and the fourth reduction (n = 4) is 0.0625 of the basic.

This series can be applied to length, area, volume, or whatever physical quantity proves to be the most advantageous. An example of the use of this series is the ISO paper sizes, which start at 1 m^2 and continue downward in size to 0.5 m^2, 0.25 m^2, and so on.

This nondecimal system is common and easily understood. Its major disadvantage is the introduction of an additional decimal place with each successive halving; this could be a problem if a wide range of product sizes was intended.

Even decimal series.—The even decimal series is a very simple sequence based on incremental changes determined by multiples or quotients of whole numbers. Examples of this series are listed in table 5-8.

Table 5-8.—*Examples of decimal series*

Multiples:	
n = 10	10, 20, 30, 40, etc.
n = 2	2, 4, 6, 8, etc.
Quotients:	
n = 10	1.0, 0.9, 0.8, 0.7, etc.
n = 5	1.0, 0.8, 0.6, 0.4, etc.
n = 2	1.0, 0.5

The selection of which series to use is determined by the beginning, the stopping point, and the desired number of increments in between.

The Renard preferred number series.—There are several Renard series; each is based on the concept of a series of numbers increasing by a fixed ratio. This method avoids the use of small increments near the upper end of a range which often occurs when the series increases by a fixed value.

The Renard R3 series is considered by many to be the ideal metric series; this is also called the 1-2-5 series. It is based on a ratio of the cube root of 10 ($\simeq$2.154); each successive number is approximately 2.154 times the preceding number. The numbers have been rounded and rationalized and the series is now written 1, 2, 5, 10, 20, 50, etc. These numbers can also be used in a denominator; this establishes the quotient series of 1.0, 0.5, 0.2, 0.1, 0.05, 0.02, etc.

The basic Renard R3 series is the one used by the ISO in establishing the preferred drafting scales. The Renard R20 and R40 series were used to establish the ISO standard sieve sizes.

Table 5-9 lists other examples of Renard series.

CHAPTER 5—UNIT AND FORMULAE CONVERSIONS 185

Table 5-9.—*Renard series*

R3	
Multiple	1, 2, 5, 10, 20, 50, 100, 200, etc.
Quotient	1.0, 0.5, 0.2, 0.1, 0.05, 0.02, 0.01, 0.005, etc.
R5	
Multiple	10, 16, 25, 40, 63, 100, 160, 250, 400, 630, etc.
R10	
Multiple	10, 12.5, 16, 20, 25, 31.5, 40, 50, 63, 80, 100, 125, 160, etc.
R20	
Multiple	10, 11.2, 12.5, 14, 16, 17.8, 20, 22.4, 25, 28.2, etc.

Note: In an R(n) series, the nth step after the first number is 10 times the first number. This is because the propagating ratio is the nth root of 10.

A Renard series is quite useful in large families of sizes. It can maintain a limit on the number of sizes and still satisfy all engineering requirements. The series can be applied to any physical quantity selected as the critical dimension of manufactured hardware; dimensions which can be graduated with a Renard series include length, volume, power output, or mass.

Linear increments.—Table 5-10 lists suggested increments to be adopted in establishing units of length. This table is recommended for use by the Bureau in the design of its construction projects.

Similar tables can be developed for area and volume, but these would not have the application that table 5-10 has.

Scientific Notation

Scientific notation is a means of expressing numbers in a fixed format. It is beneficial because very large and very small numbers can be written in the same number of spaces; the significant difference between large and small numbers is the size and sign of the exponent. Scientific notation is the format used to express the multipliers in the conversion tables.

The scientific notation format used in the conversion tables lists the coefficient[4] followed by a capital E, indicating exponent, along with a positive or negative two-digit number. The numbers represent the power of 10 by which the coefficient must be mutliplied to obtain the conversion factor.

Examples of the scientific notation are:

1.234 567	E+08 = 1.234 567 x 10^8	= 123 456 700
2.700	E+01 = 2.7 x 10^1	= 27
1.000	= 1.0	= 1
8.00	E-04 = 8.0 x 10^{-4}	= 0.0008
7.124 611	E-02 = 7.124 611 x 10^{-2}	= 0.071 246 11

---

[4]Coefficients are written with one nonzero digit to the left of the decimal and a maximum of six digits to the right of the decimal marker.

Table 5-10.—Incremental measures of length

Preference ranking	Very small dimensions up to about 50 mm	Small dimensions up to about 300 mm	Medium dimensions up to about 6000 mm	Large dimensions (site layouts, roadworks, etc.)	Other numerical values	
					Range 100 to 1000	Range 1 to 100
1st	Multiples of 10 mm	Multiples of 100 mm	Multiples of 600 mm	Multiples of 10 m	Multiples of 100	Multiples of 10
2nd	Multiples of 5 mm	Multiples of 50 mm	Multiples of 300 mm	Multiples of 5 m	Multiples of 50	Multiples of 5
3rd	Multiples of 2 mm	Multiples of 25 mm	Multiples of 100 mm	Multiples of 1 m	Multiples of 20	Multiples of 2
4th	Multiples of 1 mm	Multiples of 10 mm	Multiples of 50 mm	Multiples of 0.5 m	Multiples of 10	Multiples of 1
5th		Multiples of 5 mm	Multiples of 25 mm	Multiples of 0.1 m		
6th		Multiples of 2 mm	Multiples of 10 mm			
7th		Multiples of 1 mm				

CHAPTER 5—UNIT AND FORMULAE CONVERSIONS

The general guidelines used for the conversion tables are:
(a) No exponent listing for 10^0;
(b) Coefficients rounded to the sixth decimal position, when required;
(c) The second group of insignificant zeros is dropped.

Organization and Use of the Conversion Tables

There are 32 conversion tables listed in the following section. Two of these tables are specialized equivalency tables (Energy content of fuels and Grain conversions); these are included because of the nature of the work being accomplished by the Bureau of Reclamation. The remaining 30 tables are groupings of particular physical quantities. Several tables include subgroups, each having its own dimensional representation. An example of this is velocity; this is divided into linear and angular velocities with dimensional representations of L/T and θ/T, respectively. The common units are listed alphabetically within each group or subgroup, along with the SI unit name to which it is to be converted and the appropriate conversion factor.

Most of the tables employ the conversion factors written in the scientific notation format previously discussed; other tables use a tabular or checkerboard equivalency listing. These latter tables were prepared in this manner because it was believed that such an arrangement would be more functional and efficient for that particular information.

The tables which employ the scientific notation can also be easily used to determine relationships between any of the units listed in the table having a common SI conversion. Refer to figure 5-1 for an example. Figure 5-1 is an excerpted portion of the length (L) conversion table. Note that the nautical chain and league are highlighted; to determine how many nautical chains there are in one league, simply divide the conversion factor for league (B) by the conversion factor for the nautical chain (A). This shows that there are

$$\frac{4.828\ 032 \times 10^3}{4.5.72} = 1056 \text{ nautical chains in one league.}$$

chain, engineer's	meter (m)	*3.048	E+01	
chain, nautical	meter (m)	4.572		A
fathom	meter (m)	*1.828 800		
fermi [obsolete, replaced by femtometer]	meter (m)	*1.000	E-15	
femtometer (fm)	meter (m)	*1.000	E-15	
foot [U.S. survey] (ft)	meter (m)	3.048 006	E-01	
foot [International] (ft)	meter (m)	*3.048	E-01	
furlong (fur)	meter (m)	2.011 680	E+02	
inch (in)	meter (m)	*2.540	E-02	
kilometer (km)	meter (m)	*1.000	E+03	
league	meter (m)	4.828 032	E+03	B
link, surveyor's	meter (m)	2.011 680	E-01	

Figure 5-1.—Excerpted portion of length conversion table.

Any SI unit name which is not listed in the tables can easily be determined by the application of dimensional substitution. An example of this would be the conversion of pound foot per second (lb·ft/s) to its SI equivalent, kilogram meter per second (kg·m/s). Applying the relationships listed in the mass and length tables, 1 lb = 0.453 592 4 kg and 1 ft = 0.3048 m, substitution shows that:

$$1 \text{ lb·ft/s} = [(0.453\ 592\ 4 \text{ kg})(0.3048 \text{ m})/\text{s}] = 0.138\ 255 \text{ kg·m/s}$$

Conversion Tables

The following conversion tables are contained in this chapter:

Acceleration
Area
Chemical concentrations
Density—Mass capacity
Electricity—Magnetism
Electromagnetic radiation
Energy: Work—Thermal—Electrical
Energy content of fuels
Energy per area time
Flow
Force
Force per length
Frequency
Grain conversions
Heat
Hydraulic conductivity—Permeability

Illumination
Inertia
Length
Linear density
Load concentration
Mass
Plane angles
Power
Pressure—Stress
Temperature
Time
Torque—Bending moment
Transmissivity
Velocity—Speed
Viscosity
Volume—Capacity

CHAPTER 5—UNIT AND FORMULAE CONVERSIONS

Table 5-11.—*Acceleration*

To convert from	To	Multiply by	
Angular (θ/T^2):			
degree per second squared	radian per second squared (rad/s^2)	1.745 329	E-02
revolution per minute squared (r/min^2)	radian per second squared (rad/s^2)	1.745 329	E-03
revolution per second squared (r/s^2)	radian per second squared (rad/s^2)	6.283 185	
Linear (L/T^2):			
centimeter per second squared (cm/s^2)	meter per second squared (m/s^2)	*1.000	E-02
foot per second squared (ft/s^2)	meter per second squared (m/s^2)	*3.048	E-01
gal or Galileo (Gal)	meter per second squared (m/s^2)	*1.000	E-02
gravity, standard free fall (G or g)	meter per second squared (m/s^2)	*9.806 650	
inch per second squared (in/s^2)	meter per second squared (m/s^2)	*2.540	E-02

* Exact conversion.

Table 5-12.—Area (L^2)

To convert from	To	Multiply by	
acre [U.S. survey]	square meter (m²)	4.046 873	E+03
are	square meter (m²)	*1.000	E+02
barn	square meter (m²)	*1.000	E−28
circular mil (cmil)	square meter (m²)	5.067 075	E−10
hectare (ha)	square meter (m²)	*1.000	E+04
section [U.S. survey]	square meter (m²)	2.589 998	E+06
square centimeter (cm²)	square meter (m²)	*1.000	E−04
square chain	square meter (m²)	*1.562 500	E−04
square foot [International] (ft²)	square meter (m²)	*9.290 304	E−02
square foot [U.S. survey] (ft²)	square meter (m²)	9.290 341	E−02
square inch (in²)	square meter (m²)	*6.451 600	E−04
square kilometer (km²)	square meter (m²)	*1.000	E+06
square mile [International] (mi²)	square meter (m²)	2.589 988	E+06
square mile [U.S. survey] (mi²)	square meter (m²)	2.589 998	E+06
square rod [U.S. survey] (rod²)	square meter (m²)	2.529 295	E+01
square yard (yd²)	square meter (m²)	8.361 274	E−01
township	square meter (m²)	9.323 993	E+07

* Exact conversion.

CHAPTER 5—UNIT AND FORMULAE CONVERSIONS

Table 5-13.—*Chemical concentration*

A — Mass percent of solute
B — Molecular mass of solvent (g)
E — Molecular mass of solute (g)
F — Milligrams of solute per liter of solution (mg/L)
G — Molality (mol/kg)
M — Molarity (mol/L)
N — Mole fraction
R — Solution density (kg/L)
X — Part per million — mass basis
Y — Part per million — volume basis
T — Normality, equivalent per liter (eq/L)
U — Equivalent mass of solute ion

Concentration of solute SOUGHT	Concentration of solute GIVEN							
	A: Percent (mass)	N: Mole fraction	G: Molality (mol/kg)	M: Molarity[a] (mol/L)	X: Part per million[b] (mass)	Y: Part per million[c] (vol.)	F: Solute concentration (mg/L)	T: Normality (eq/L)
A: percentage number	—	$\dfrac{100\,EN}{EN + B(1-N)}$	$\dfrac{100\,EG}{1000 + EG}$	$\dfrac{EM}{10R}$	$10^{-4}\,X$		$\dfrac{F}{10^{4}\,R}$	
N: decimal number	$\dfrac{(A/E)}{(A/E) + \dfrac{100-A}{B}}$	—	$\dfrac{BG}{BG + 1000}$	$\dfrac{BM}{M(B-E) + 10^{3}\,R}$			$\dfrac{10^{-3}\,BF}{F(B-E) + 10^{3}\,R}$	
F: mg/L	$10^{4}\,AR$	$\dfrac{10^{6}\,ENR}{EN + (1-N)B}$	$\dfrac{10^{6}\,EGR}{1000 + EG}$	$10^{3}\,EM$	RX^{d}	RY^{e}	—	$10^{3}\,UT$
G: mol/kg	$\dfrac{10^{3}\,A}{(100-A)E}$	$\dfrac{10^{3}\,N}{B(1-N)}$	—	$\dfrac{10^{3}\,M}{10^{3}\,R - EM}$			$\dfrac{F}{E(10^{3}\,R - F)}$	
M: mol/L	$\dfrac{10\,AR}{E}$	$\dfrac{10^{3}\,NR}{EN + B(1-N)}$	$\dfrac{10^{3}\,GR}{1000 + EG}$	—			$\dfrac{F}{10^{3}\,E}$	

[a] Customarily used for liquids, especially aqueous solutions.
[b] Customarily used primarily for solid and liquid materials.'
[c] Customarily used for gases, especially polluted air.
[d] In most acqueous solutions (but not every case), the unit of parts per million (p/m), mass basis, has been erroneously attached to values which were actually determined in milligrams per liter (mg/L). For such cases, the conversion factor is 1.000. For moderate and dilute saline solutions (less than 10 000 mg/L total dissolved solids), the use of a conversion factor of 1.000 gives an error which is usually less than 1 percent; at higher concentrations, such as in brines, the error becomes significant and the correct solution density needs to be incorporated.
[e] More commonly expressed in micrograms per cubic meter ($\mu g/m^{3}$); to convert from milligrams per liter to micrograms per cubic meter, multiply by 10^{6}. The density value depends upon the prevailing temperature and pressure; this relationship assumes Ideal Gas or Amagat's Law holds.

Table 5-14.—*Density—Mass capacity* (M/L^3)

To convert from	To	Multiply by	
gram per cubic centimeter (g/cm^3)	kilogram per cubic meter (kg/m^3)	*1.000	E+03
gram per liter (g/L)	kilogram per cubic meter (kg/m^3)	*1.000	E+06
megagram per cubic meter (Mg/m^3)	kilogram per cubic meter (kg/m^3)	*1.000	E+03
metric ton per cubic meter (t/m^3)	kilogram per cubic meter (kg/m^3)	*1.000	E+03
milligram per liter (mg/L)	kilogram per cubic meter (kg/m^3)	*1.000	E+03
ounce per cubic inch (oz/in^3)	kilogram per cubic meter (kg/m^3)	1.729 994	E+03
ounce per gallon (oz/gal)	kilogram per cubic meter (kg/m^3)	7.489 152	
ounce per pint (oz/pt)	kilogram per cubic meter (kg/m^3)	9.361 440	E-01
pound per cubic inch (lb/in^3)	kilogram per cubic meter (kg/m^3)	2.767 990	E+04
pound per cubic foot (lb/ft^3)	kilogram per cubic meter (kg/m^3)	1.601 846	E+01
pound per cubic yard (lb/yd^3)	kilogram per cubic meter (kg/m^3)	5.932 764	E-01
pound per gallon (lb/gal)	kilogram per cubic meter (kg/m^3)	1.198 264	E+02
slug per cubic foot $(slug/ft^3)$	kilogram per cubic meter (kg/m^3)	5.153 788	E+02
ton [short] per cubic yard (ton/yd^3)	kilogram per cubic meter (kg/m^3)	1.186 553	E+03

* Exact conversion.

Table 5-15.—Electricity—Magnetism

To convert from[a,b]	To	Multiply by	
Capacitance:			
abfarad	farad (F)	*1.000	E+09
coulomb per volt (C/V)	farad (F)	*1.000	
EMU of capacitance	farad (F)	*1.000	E+09
ESU of capacitance	farad (F)	*1.112 650	E-12
farad (F)	farad (F)	*1.000	
statfarad	farad (F)	*1.112 650	E-12
Conductance:			
abmho	siemens (S)	*1.000	E+09
mho (Ω^{-1})	siemens (S)	*1.000	
statmho	siemens (S)	*1.112 650	E-12
Current:			
ampere (A)	ampere (A)	*1.000	
EMU of current	ampere (A)	*1.000	E+01
ESU of current	ampere (A)	*3.335 640	E-10
gilbert	ampere (A)	7.957 747	E-01
statampere	ampere (A)	3.335 640	E-10
Current per length:			
milliampere per centimeter (mA/cm)	ampere per meter (A/m)	*1.000	E-01
oersted	ampere per meter (A/m)	7.957 747	E+01
Electric charge:			
abcoulomb	coulomb (C)	*1.000	E+01
ampere-hour (A·h)	coulomb (C)	*3.600	E+03
coulomb (C)	coulomb (C)	*1.000	
faraday [based on ^{12}C]	coulomb (C)	*9.648 700	E+04
statcoulomb	coulomb (C)	*3.335 640	E-10
Inductance:			
abhenry	henry (H)	*1.000	E-09
EMU of inductance	henry (H)	*1.000	E-09
ESU of inductance	henry (H)	8.987 554	E+11
henry (H)	henry (H)	*1.000	
stathenry	henry (H)	8.987 554	E+11

Table 5-15.—*Electricity—Magnetism*—Continued

To convert from[a,b]	To	Multiply by	
Magnetic polarization and flux density:			
gamma	tesla (T)	*1.000	E−09
gauss	tesla (T)	*1.000	E−04
Magnetic flux:			
maxwell	weber (Wb)	*1.000	E−08
unit pole	weber (Wb)	1.256 637	E−07
weber (Wb)	weber (Wb)	*1.000	
Electric potential:			
abvolt	volt (V)	*1.000	E−08
EMU of electric potential	volt (V)	*1.000	E−08
ESU of electric potential	volt (V)	2.997 925	E+02
statvolt	volt (V)	2.997 925	E+02
volt (V)	volt (V)	*1.000	
Resistance:			
EMU of resistance	ohm (Ω)	*1.000	E−09
ESU of resistance	ohm (Ω)	8.987 554	E+11
ohm (Ω)	ohm (Ω)	*1.000	
statohm	ohm (Ω)	8.987 554	E+11
Resistivity:			
ohm centimeter ($\Omega \cdot$cm)	ohm meter ($\Omega \cdot$m)	*1.000	E−02
ohm foot ($\Omega \cdot$ft)	ohm meter ($\Omega \cdot$m)	*3.048	E−01
ohm meter ($\Omega \cdot$m)	ohm meter ($\Omega \cdot$m)	*1.000	
ohm circular-mil per foot ($\Omega \cdot$cmil/ft)	ohm square millimeter per meter ($\Omega \cdot$mm^2/m)	1.662 426	E−03

* Exact conversion.
[a] EMU—electromagnetic unit.
[b] ESU—electrostatic unit.

CHAPTER 5—UNIT AND FORMULAE CONVERSIONS

Table 5-16.—*Electromagnetic radiation*

To convert from	To	Multiply by
radiation absorbed dose (rad)	gray (Gy)	1.000 E-02
joule per kilogram (J/kg)	gray (Gy)	1.000
roentgen equivalent man (rem)	gray (Gy)	See footnote b
erg per gram (erg/g)	gray (Gy)	1.000 E-04
roentgen equivalent physical (rep)	coulomb per kilogram[a] (C/kg)	2.580 E-04
roentgen (R)	coulomb per kilogram[a] (C/kg)	2.580 E-04
gram-rad	joule (J)	1.000 E-05
gram-roentgen	joule (J)	8.690 E-06
langley	joule per square meter (J/m^2)	4.184 E+04
calorie [IT] per square centimeter[c] (cal$_{IT}$/cm^2)	joule per square meter (J/m^2)	4.186 800 E+04
pyron	watt per square meter (W/m^2)	6.978 E+02
curie (Ci)	becquerel (Bq)	3.700 E+10
rutherford (rd)	becquerel (Bq)	1.000 E+06
disintegration per second	becquerel (Bq)	1.000

[a] Can be further converted to grays only with knowledge of the radiation media. Example: In air, 1 R equals 8.69 x 10^{-3} Gy.
[b] Conversion factor depends on the absorbed dose in rads and modifying factors.
[c] One reference [13] identified the langley and the cal$_{IT}$/cm^2 as equivalents; unable to determine discrepancy with other publications.

Table 5-17.—*Energy: Work—Thermal—Electrical* (ML/T^2)

To convert from	To	Multiply by	
British thermal unit $[IT]$ [a] (Btu_{IT})	joule (J)	1.055 056	E+03
British thermal unit $[tc]$ [a] (Btu_{tc})	joule (J)	1.054 350	E+03
British thermal unit [mean] (Btu_m)	joule (J)	1.055 870	E+03
calorie $[IT]$ (cal_{IT})	joule (J)	*4.186 800	
calorie $[tc]$ (cal_{tc})	joule (J)	*4.184	
calorie [mean] (cal_m)	joule (J)	4.190 020	
electronvolt (eV)	joule (J)	1.062 190	E−19
erg	joule (J)	*1.000	E−07
foot-pound force (ft·lbf)	joule (J)	1.355 818	
foot-poundal (ft·pdl)	joule (J)	4.214 011	E−02
kilowatt hour (kW·h)	joule (J)	*3.600	E+06
quad [quadrillion Btu_{IT}]	joule (J)	1.055 056	E+18
therm	joule (J)	1.055 056	E+08
ton [TNT equivalent]	joule (J)	4.184	E+09[b]
watt hour (W·h)	joule (J)	*3.600	E+03
watt second (W·s)	joule (J)	*1.000	

* Exact conversion.
[a] The IT and tc symbols represent International Table and thermochemical, respectively.
[b] This is a defined value, not measured. Other numerical conversion factors, all approximately equal, were also found.

Table 5-18.—*Energy content of fuels*

Fuel	Unit of measure	Joules per unit[a]
Coal:		
anthracite	metric ton (0.65 m^3)	29 524.0 x 10^6
bituminous	metric ton (0.75 m^3)	28 731.0 x 10^6
average	metric ton (0.80 m^3)	27 937.0 x 10^6
Gas:		
natural	cubic meter[b]	38.4 x 10^6
liquified petroleum[c]	cubic meter[b]	93.9 x 10^6
Petroleum:		
crude	cubic meter[d]	37.1 x 10^9
aviation gasoline	liter	34.8 x 10^6
automobile gasoline	liter	34.8 x 10^6
diesel oil	liter	38.7 x 10^6
distillate fuel oil, #2	liter	38.7 x 10^6
kerosene	liter	37.6 x 10^6
jet fuel, naphtha type	liter	35.4 x 10^6

[a] Approximate values.
[b] Density at STP.
[c] Includes butane and propane.
[d] A cubic meter of crude oil has a mass of 855 kg (approximately).

Table 5-19.—*Energy per area time* (M/T^3)

To convert from	To	Multiply by	
British thermal unit [tc] per square foot second ($Btu_{tc}/(ft^2 \cdot s)$)	watt per square meter (W/m^2)	1.134 893	E+04
British thermal unit [tc] per square foot minute ($Btu_{tc}/(ft^2 \cdot min)$)	watt per square meter (W/m^2)	1.891 489	E+02
British thermal unit [tc] per square foot hour ($Btu_{tc}/(ft^2 \cdot h)$)	watt per square meter (W/m^2)	3.152 481	
British thermal unit [tc] per second ($Btu_{tc}/(in^2 \cdot s)$)	watt per square meter (W/m^2)	1.634 246	E+06
calorie [tc] per square centimeter minute ($cal_{tc}/(cm^2 \cdot min)$)	watt per square meter (W/m^2)	6.973 333	E+02
erg per square centimeter second ($erg/(cm^2 \cdot s)$)	watt per square meter (W/m^2)	*1.000	E−06
watt per square centimeter (W/cm^2)	watt per square meter (W/m^2)	*1.000	E+04
watt per square inch (W/in^2)	watt per square meter (W/m^2)	1.550 003	E+03

* Exact conversion.

CHAPTER 5—UNIT AND FORMULAE CONVERSIONS

Table 5-20.—Flow

To convert from	To	Multiply by	
Mass flow (M/T):			
pound per second (lb/s)	kilogram per second (kg/s)	4.535 924	E-01
pound per minute (lb/min)	kilogram per second (kg/s)	7.559 873	E-03
pound per hour (lb/h)	kilogram per second (kg/s)	1.259 979	E-04
slug per second (slug/s)	kilogram per second (kg/s)	1.459 390	E+01
ton per hour (ton/h)	kilogram per second (kg/s)	2.519 958	E-01
Volume flow (L^3/T):			
acre-foot per day (acre-ft/d)	cubic meter per second (m^3/s)	1.427 641	E-02
cubic dekameter per day (dam^3/d)	cubic meter per second (m^3/s)	1.157 407	E-02
cubic foot per second (ft^3/s)	cubic meter per second (m^3/s)	2.831 685	E-02
cubic foot per minute (m^3/min)	cubic meter per second (m^3/s)	4.719 474	E-04
cubic meter per minute (m^3/min)	cubic meter per second (m^3/s)	1.666 667	E-02
cubic meter per day (m^3/d)	cubic meter per second (m^3/s)	1.157 407	E-05
cumec	cubic meter per second (m^3/s)	*1.000	
cusec	cubic meter per second (m^3/s)	2.831 685	E-02
gallon per second (gal/s)	cubic meter per second (m^3/s)	3.785 412	E-03
gallon per minute (gal/min)	cubic meter per second (m^3/s)	6.309 020	E-05
liter per second (L/s)	cubic meter per second (m^3/s)	*1.000	E-03
liter per minute (L/min)	cubic meter per second (m^3/s)	1.666 667	E-05

Table 20.—*Flow*—Continued

To convert	To	Multiply by
Volume flow: (continued)		
milliliter per second (mL/s)	cubic meter per second (m^3/s)	*1.000 E−06
million gallons per day (10^6 gal/d)	cubic meter per second (m^3/s)	4.381 260 E−02
miner's inch[a]	cubic meter per second (m^3/s)	3.277 413 E−07
miner's inch[b]	cubic meter per second (m^3/s)	4.096 766 E−07
miner's inch[c]	cubic meter per second (m^3/s)	4.267 465 E−07
Miscellaneous:		
gallon per horsepower-hour (gal/(hp·h))	cubic meter per joule (m^3/J)	1.410 089 E−09
mile per gallon (mi/gal)	liter per 100 km (L/100 km)	See footnote d
pound [mass] per horsepower-hour (lb/(hp·h))	kilogram per joule (kg/J)	1.689 659 E−07
Water vapor transmission:		
permeance at 0 °C [perm 0 °C)]	kilogram per pascal second square meter (kg/(Pa·s·m^2))	5.721 350 E−11
permeance at 23 °C [perm (23 °C)]	kilogram per pascal second square meter (kg/(Pa·s·m^2))	5.745 250 E−11
permeability at 0 °C [perm·in (0 °C)]	kilogram per pascal second square meter (kg/(Pa·s·m^2))	1.453 220 E−12
permeability at 23 °C [perm·in (23 °C)]	kilogram per pascal second square meter (kg/(Pa·s·m^2))	1.459 290 E−12

* Exact conversion.
[a] Applies to Idaho, Kansas, Nebraska, New Mexico, North Dakota, South Dakota, Utah, Washington, and southern California.
[b] Applies to Arizona, Montana, Nevada, Oregon, and northern California.
[c] Applies to Colorado.
[d] Do not multiply by factor; divide 235.215 by the mile per gallon value.

CHAPTER 5—UNIT AND FORMULAE CONVERSIONS

Table 5-21.—*Force* (ML/T^2)

To convert from	To	Multiply by	
crinal	newton (N)	*1.000	E-01
dyne (dyn)	newton (N)	*1.000	E-05
kilogram force (kgf)	newton (N)	*9.806 650	
kilopond	newton (N)	*9.806 650	
kip	newton (N)	4.448 222	E+03
ounce force	newton (N)	2.780 139	E-01
pound force (lbf)	newton (N)	4.448 222	
poundal (pdl)	newton (N)	1.382 550	E-01
ton force	newton (N)	8.896 444	E+03

* Exact conversion.

Table 5-22.—*Force per length* (M/T^2)

To convert from	To	Multiply by	
dyne per centimeter (dyn/cm)	newton per meter (N/m)	*1.000	E-03
kilogram force per meter (kgf/m)	newton per meter (N/m)	*9.806 650	
pound per foot (lb/ft)	newton per meter (N/m)	1.459 390	E+01
pound per inch (lb/in)	newton per meter (N/m)	1.751 268	E+02

* Exact conversion.

Table 5-23.—*Frequency* (T^{-1})

To convert from	To	Multiply by	
cycle per hour (c/h)	hertz (Hz)	2.777 778	E-04
cycle per minute (c/min)	hertz (Hz)	1.666 667	E-02
cycle per second (c/s)	hertz (Hz)	*1.000	
fresnel	hertz (Hz)	*1.000	E+12

* Exact conversion.

Table 5-24.—*Grain conversions*

	Pounds per bushel	Kilograms per cubic meter	Bushels per metric ton[a]
Barley	48	617.8	45.9
Corn	56	720.8	39.4
Oats	32	411.9	68.9
Rye	56	720.8	39.4
Soybeans	60	772.3	36.7
Wheat	60	772.3	36.7
Rice[b]	45	579.2	49.0
Sorghum	56	720.8	39.4

[a] To obtain metric tons per hectare, divide bushels per acre by the product of this column and 0.4047. Example: A corn harvest of 110 bu/acre equals 6.9 metric tons per hectare (t/ha), [110 ÷ (39.4)(0.4047)] = 6.9.
[b] Applies to rough rice from field.
Note: Storage capacity for grains will be expressed in cubic meters.

CHAPTER 5—UNIT AND FORMULAE CONVERSIONS

Table 5-25.—*Heat*

To convert from[a]	To	Multiply by	
Absorption per area:			
[time dependent]			
calorie [tc] per square centimeter second ($cal_{tc}/(cm^2 \cdot s)$)	watt per square meter (W/m^2)	*4.184	E+04
calorie [tc] per square centimeter minute ($cal_{tc}/(cm^2 \cdot min)$)	watt per square meter (W/m^2)	6.973 333	E+02
[time independent]			
British thermal unit [IT] per square foot (Btu_{IT}/ft^2)	joule per square meter (J/m^2)	1.135 653	E+04
British thermal unit [tc] per square foot (Btu_{tc}/ft^2)	joule per square meter (J/m^2)	1.134 893	E+04
calorie [tc] per square centimeter (cal_{tc}/cm^2)	joule per square meter (J/m^2)	*4.184	E+04
Flow rate:			
British thermal unit [mean] per hour (Btu_m/h)	watt (W)	9.084 924	E+05
British thermal unit [IT] per hour (Btu_{IT}/h)	watt (W)	9.077 920	E+05
calorie [tc] per minute (cal_{tc}/min)	watt (W)	6.973 333	E−02
calorie [tc] per second (cal_{tc}/s)	watt (W)	*4.184	
calorie [mean] per minute (cal_m/min)	watt (W)	6.983 367	E−02
joule per second (J/s)	watt (W)	*1.000	
Quantity:			
British thermal unit (Btu) calorie (cal) joule (J) therm (therm)	See footnote b		

Table 5-25.—Heat—Continued

To convert from[a]	To	Multiply by
Specific heat capacity and specific entropy:		
British thermal unit [IT] per pound [mass] degree Fahrenheit ($\text{Btu}_{IT}/(\text{lb}\cdot{}^\circ\text{F})$)	joule per kilogram kelvin (J/(kg·K))	*4.186 800 E+03
British thermal unit [tc] per pound [mass] degree Fahrenheit ($\text{Btu}_{tc}/(\text{lb}\cdot{}^\circ\text{F})$)	joule per kilogram kelvin (J/(kg·K))	*4.184 E+03
calorie [tc] per gram degree Celsius ($\text{cal}_{tc}/(\text{g}\cdot{}^\circ\text{C})$)	joule per kilogram kelvin [J/(kg·K)]	*4.184 E+03
calorie [IT] per gram degree Celsius ($\text{cal}_{IT}/(\text{g}\cdot{}^\circ\text{C})$)	joule per kilogram kelvin (J/(kg·K))	*4.186 800 E+03
Specific latent heat or specific energy:		
British thermal unit [IT] per pound [mass] ($\text{Btu}_{IT}/\text{lb}$)	joule per kilogram (J/kg)	*2.326 E+03
British thermal unit [tc] per pound [mass] ($\text{Btu}_{tc}/\text{lb}$)	joule per kilogram (J/kg)	2.324 444 E+03
calorie [tc] per gram (cal_{tc}/g)	joule per kilogram (J/kg)	*4.184 E+03
calorie [IT] per gram (cal_{IT}/g)	joule per kilogram (J/kg)	*4.186 800 E+03
Thermal conductance or coefficient of heat transfer:		
British thermal unit [IT] per second square foot degree Fahrenheit ($\text{Btu}_{IT}/(\text{s}\cdot\text{ft}^2\cdot{}^\circ\text{F})$)	watt per square meter kelvin (W/(m^2·K))	2.044 175 E+04
British thermal unit [tc] per second square foot degree Fahrenheit ($\text{Btu}_{tc}/(\text{s}\cdot\text{ft}^2\cdot{}^\circ\text{F})$)	watt per square meter kelvin (W/(m^2·K))	2.042 808 E+04

CHAPTER 5—UNIT AND FORMULAE CONVERSIONS 205

Table 5-25.—Heat—Continued

To convert from[a]	To	Multiply by	
Thermal conductance or coefficient of heat transfer: (Continued)			
British thermal unit $[IT]$ per hour square foot degree Fahrenheit $(Btu_{IT}/(h \cdot ft^2 \cdot {}^\circ F))$	watt per square meter kelvin $(W/(m^2 \cdot K))$	5.678 263	
British thermal unit $[tc]$ per hour square foot degree Fahrenheit $(Btu_{tc}/(h \cdot ft^2 \cdot {}^\circ F))$	watt per square meter kelvin $(W/(m^2 \cdot K))$	5.674 466	
Thermal conductivity:			
British thermal unit $[IT]$-foot per hour square foot degree Fahrenheit $(Btu_{IT} \cdot ft/(h \cdot ft^2 \cdot {}^\circ F))$	watt per meter kelvin $(W/(m \cdot K))$	1.730 735	
British thermal unit $[tc]$-foot per hour square foot degree Fahrenheit $(Btu_{tc} \cdot ft/(h \cdot ft^2 \cdot {}^\circ F))$	watt per meter kelvin $(W/(m \cdot K))$	1.729 577	
British thermal unit $[IT]$-inch per hour square foot degree Fahrenheit $(Btu_{IT} \cdot in/(h \cdot ft^2 \cdot {}^\circ F))$	watt per meter kelvin $(W/(m \cdot K))$	1.442 279	E-01
British thermal unit $[tc]$-inch per hour square foot degree Fahrenheit $(Btu_{tc} \cdot in/(h \cdot ft^2 \cdot {}^\circ F))$	watt per meter kelvin $[W/(m \cdot K)]$	1.441 314	E-01
British thermal unit $[IT]$-inch per second square foot degree Fahrenheit $(Btu_{IT} \cdot in/(s \cdot ft^2 \cdot {}^\circ F))$	watt per meter kelvin $(W/(m \cdot K))$	5.192 204	E+02

Table 5-25.—*Heat*—Continued

To convert from[a]	To	Multiply by
Thermal conductivity: (Continued)		
British thermal unit [tc]-inch per second square foot degree Fahrenheit ($Btu_{tc} \cdot in/s \cdot ft^2 \cdot °F$))	watt per meter kelvin (W/(m·k))	5.188 732 E+02
calorie [tc] per centimeter second degree Celsius ($cal_{tc}/(cm \cdot s \cdot °C)$)	watt per meter kelvin (W/(m·K))	*4.184 E+02
Thermal diffusivity:		
square foot per hour (ft^2/h)	square meter per second (m^2/s)	2.580 640 E−05
Thermal resistance:		
clo (clo)	kelvin square meter per watt (K·m²/W)	2.003 712 E−01
degree Fahrenheit hour square foot per British thermal unit [IT] ($°F \cdot h \cdot ft^2/Btu_{IT}$)	kelvin square meter per watt (K·m²/W)	1.761 102 E−01
degree Fahrenheit hour square foot per British thermal unit [tc] ($°F \cdot h \cdot ft^2/Btu_{tc}$)	kelvin square meter per watt (K·m²/W)	1.762 280 E−01

* Exact conversion.
[a] In this listing, the subscripts IT, tc, and m designate International Table, thermochemical, and mean, respectively.
[b] Listed for reference only; conversions involving these units are in the energy table.

Table 5-26.—*Hydraulic conductivity—Permeability* $(L^3/L^2 T)$

To convert from	To	Multiply by	
cubic foot per square foot day[a] $(ft^3/(ft^2 \cdot d))$	cubic meter per square meter second[b] $(m^3/(m^2 \cdot s))$	3.527 778	E-06
cubic meter per square meter day[c] $(m^3/(m^2 \cdot d))$	cubic meter per square meter second $(m^3/(m^2 \cdot s))$	1.157 407	E-05
cubic meter per square meter year[d] $(m^3/(m^2 \cdot a))$	cubic meter per square meter second $(m^3/(m^2 \cdot s))$	3.170 979	E-08
darcy	cubic meter per square meter second $(m^3/(m^2 \cdot s))$	8.574	E-06
gallon per square foot day $(gal/(ft^2 \cdot d))$	cubic meter per square meter second $(m^3/(m^2 \cdot s))$	4.715 937	E-07
liter per square meter day $(L/(m^2 \cdot d))$	cubic meter per square meter second $(m^3/(m^2 \cdot s))$	1.157 407	E-08

[a] Also written as foot per day (ft/d).
[b] Also written as meter per second (m/s).
[c] Also written as meter per day (m/d).
[d] Also written as meter per year (m/a).

Table 5-27.—*Illumination*

To convert from	To	Multiply by	
Luminous intensity:			
candle, international standard	candela (cd)	1.019	
carcel	candela (cd)	9.789 500	
new candle	candela (cd)	*1.000	
Luminous flux:			
candela per steradian (cd/sr)	lumen (lm)	*1.000	
Luminous energy:			
lumberg	lumen second (lm·s)	*1.000	
talbot	lumen second (lm·s)	*1.000	
Luminance:			
apostilb (asb)	candela per square meter (cd/m^2)	3.183 098	E−01
candela per square foot (cd/ft^2)	candela per square meter (cd/m^2)	1.076 391	E+01
footcandle, equivalent	candela per square meter (cd/m^2)	3.426 259	
footlambert (fL)	candela per square meter (cd/m^2)	3.426 259	
lambert (L)	candela per square meter (cd/m^2)	3.183 099	E+03
nit (nt)	candela per square meter (cd/m^2)	*1.000	
stilb (sb)	candela per square meter (cd/m^2)	*1.000	E+04
Intensity of illumination:			
centimeter-candle (cm·c)	lux (lx)	*1.000	E+04
footcandle (fc)	lux (lx)	1.076 391	E+01
lumen per square centimeter (lm/cm^2)	lux (lx)	*1.000	E+04
lumen per square foot (lm/ft^2)	lux (lx)	9.290 304	E−02
lumen per square meter (lm/m^2)	lux (lx)	*1.000	

Table 5-27.—*Illumination*—Continued

To convert from	To	Multiply by	
meter candle (m·c)	lux (lx)	*1.000	
nox	lux (lx)	*1.000	E-03
phot (ph)	lux (lx)	*1.000	E+04

* Exact conversion.

Table 5-28.—*Inertia*

To convert from	To	Multiply by	
Area (L^4):			
inch^4 (in^4)	meter^4 (m^4)	4.162 314	E-08
foot^4 (ft^4)	meter^4 (m^4)	8.630 975	E-03
millimeter^4 (mm^4)	meter^4 (m^4)	*1.000	E-12
Line (L^3):			
inch cubed (in^3)	meter cubed (m^3)	1.638 706	E-05
foot cubed (ft^3)	meter cubed (m^3)	2.831 685	E-02
millimeter cubed (mm^3)	meter cubed (m^3)	*1.000	E-09
Mass (ML^2):			
pound foot squared (lb·ft^2)	kilogram meter squared (kg·m^2)	4.214 011	E-02
pound inch squared (lb·in^2)	kilogram meter squared (kg·m^2)	2.926 397	E-04
slug foot squared (slug·ft^2)	kilogram meter squared (kg·m^2)	1.355 818	

* Exact conversion.

Table 5-29.—Length (L)

To convert from	To	Multiply by	
angstrom unit (Å)	meter (m)	*1.000	E-10
astronomical unit (AU)	meter (m)	1.496	E+11[a]
caliber	meter (m)	*2.540	E-02
centimeter (cm)	meter (m)	*1.000	E-02
chain, surveyor's	meter (m)	2.011 680	E+01
chain, engineer's	meter (m)	*3.048	E+01
chain, nautical	meter (m)	4.572	
fathom	meter (m)	*1.828 800	
fermi [obsolete replaced by femtometer]	meter (m)	*1.000	E-15
femtometer (fm)	meter (m)	*1.000	E-15
foot [U.S. survey] (ft)	meter (m)	3.048 006	E-01
foot [International] (ft)	meter (m)	*3.048	E-01
furlong (fur)	meter (m)	2.011 680	E+02
inch (in)	meter (m)	*2.540	E-02
kilometer (km)	meter (m)	*1.000	E+03
league	meter (m)	4.828 032	E-03
link, surveyor's	meter (m)	2.011 680	E-01
light year (ly)	meter (m)	9.460 900	E+15[b]
microinch (μin)	meter (m)	*2.540	E-08
micrometer (μm)	meter (m)	*1.000	E-06
micron [obsolete, replaced by micrometer]	meter (m)	*1.000	E-06
mil (mil)	meter (m)	*2.540	E-05
mile [International] (mi)	meter (m)	1.609 344	E+03
mile [Statute] (mi)	meter (m)	1.609 300	E+03
mile [U.S. survey] (mi)	meter (m)	1.609 347	E+03
nautical mile (nmi)	meter (m)	*1.852	E+03
parsec	meter (m)	3.085 678	E+16
pica, printer's	meter (m)	4.217 518	E-03
point, printer's	meter (m)	3.514 598	E-04
rod	meter (m)	5.029 210	
spat	meter (m)	*1.000	E+12
yard (yd)	meter (m)	*9.144	E-01

* Exact conversion.
[a] The AU measure given was listed by two sources [13, 37]. ASTM E-380[E] [32] provides another number.
[b] The value presented is based on the speed of light determined by Dr. G. S. Simkin [28] and the sidereal year.

Table 5-30.—Linear density (M/L)

To convert from	To	Multiply by	
denier	kilogram per meter (kg/m)	1.111 111	E−07
pound per foot (lb/ft)	kilogram per meter (kg/m)	1.488 164	
pound per inch (lb/in)	kilogram per meter (kg/m)	1.785 797	E+01
tex	kilogram per meter (kg/m)	*1.000	E−06

* Exact conversion.

Table 5-31.—Load concentration (M/L^2)

To convert from	To	Multiply by	
gram per square centimeter (g/cm^2)	kilogram per square meter (kg/m^2)	*1.000	E+01
megagram per square meter (Mg/m^2)	kilogram per square meter (kg/m^2)	*1.000	E+03
metric ton per square meter (t/m^2)	kilogram per square meter (kg/m^2)	*1.000	E+03
ounce per square inch (oz/in^2)	kilogram per square meter (kg/m^2)	2.119 109	E−03
ounce per square foot (oz/ft^2)	kilogram per square meter (kg/m^2)	3.051 517	E−01
ounce per square yard (oz/yd^2)	kilogram per square meter (kg/m^2)	3.390 575	E−02
pound per square inch (lb/in^2)	kilogram per square meter (kg/m^2)	7.030 696	E+02
pound per square foot (lb/ft^2)	kilogram per square meter (kg/m^2)	4.882 428	
pound per square yard (lb/yd^2)	kilogram per square meter (kg/m^2)	5.424 920	E−01
ton per square foot (ton/ft^2)	kilogram per square meter (kg/m^2)	9.071 847	E+02

* Exact conversion.

Table 5-32.—Mass (M)

To convert from	To	Multiply by	
barrel of cement [376 lb]	kilogram (kg)	1.705 507	E+02
carat [metric]	kilogram (kg)	*2.000	E-04
carat (kt)	kilogram (kg)	2.591 956	E-04
cental	kilogram (kg)	4.535 924	E+01
centner	kilogram (kg)	4.535 924	E+01
centner [metric]	kilogram (kg)	1.000	E+02[a]
grain	kilogram (kg)	6.479 891	E-05
gram (g)	kilogram (kg)	*1.000	E-03
hundredweight [gross or long] (cwt)	kilogram (kg)	5.080 235	E+01
hundredweight [net or short] (cwt)	kilogram (kg)	4.535 924	E+01
kilogram force—second squared per meter (kgf·s^2/m)	kilogram (kg)	*9.806 650	
kilotonne (kt)	kilogram (kg)	*1.000	E+06
ounce [avoirdupois] (oz)	kilogram (kg)	2.834 952	E-02
ounce [troy/apothecary] (oz)	kilogram (kg)	3.110 348	E-02
megagram (Mg)	kilogram (kg)	*1.000	E+03
metric grain	kilogram (kg)	*5.000	E-05
metric ton (t)	kilogram (kg)	*1.000	E+03
milligram (mg)	kilogram (kg)	*1.000	E-03
pennyweight	kilogram (kg)	1.555 174	E-03
pound [avoirdupois] (lb)	kilogram (kg)	4.535 924	E-01
pound [troy/apothecary]	kilogram (kg)	3.732 417	E-01
quintal	kilogram (kg)	*1.000	E+02
sack of cement [94 lbs]	kilogram (kg)	4.263 767	E+01
slug	kilogram (kg)	1.459 390	E+01
ton [assay]	kilogram (kg)	2.916 667	E-02
ton [long]	kilogram (kg)	1.016 047	E+03
ton [short]	kilogram (kg)	9.071 847	E+02
tonne (t)	kilogram (kg)	*1.000	E+03

* Exact conversion.
[a] European metric centner is 50 percent of this value; conversion factor presented applies to the centner as used in the U.S.S.R.

CHAPTER 5—UNIT AND FORMULAE CONVERSIONS

Table 5-33.—Plane angles (θ)

	radian(s)	degree(s)	minute(s)	second(s)	degree(s) minute(s) second(s)
1 radian (rad) =	1.0	57.295 779 51	3437.7468	206 264.8	57°17'44.806"
1 degree (°) =	$17.453\ 293 \times 10^{-3}$	1.0	60.0	3600.0	1°0'0"
1 minute (') =	$29.088\ 822 \times 10^{-5}$	[b] $16.666\ 667 \times 10^{-3}$	1.0	60.0	0°1'0"
1 second (") =	$48.481\ 370 \times 10^{-7}$	[b] $27.777\ 778 \times 10^{-5}$	$16.666\ 667 \times 10^{-3}$	1.0	0°0'1"
1 grad[a] (9) =	$15.707\ 964 \times 10^{-3}$	0.9	54	3240	0°54'0"

1 sign = 30°	1 sextant = 60°	1 turn = 360°
1 octant = 45°	1 quadrant = 90°	

[a] Also spelled grade. An alternative symbol is gon.
[b] The use of the degree symbol (°) is not recommended with scientific notation.

Table 5-34.—Power (ML^2/T^3)

To convert from	To	Multiply by	
British thermal unit [IT] per hour (Btu$_{IT}$/h)	watt (W)	2.930 711	E-01
British thermal unit [tc] per hour (Btu$_{tc}$/h)	watt (W)	2.928 751	E-01
British thermal unit [tc] per minute (Btu$_{tc}$/min)	watt (W)	1.757 250	E+01
British thermal unit [tc] per second (Btu$_{tc}$/s)	watt (W)	1.054 350	E+03
calorie [tc] per minute (cal$_{tc}$/min)	watt (W)	6.973 333	E-02
calorie [tc] per second (cal$_{tc}$/s)	watt (W)	*4.184	
erg per second (erg/s)	watt (W)	*1.000	E-07
foot-pound per hour (ft·lb/h)	watt (W)	3.766 161	E-04
foot-pound per minute (ft·lb/min)	watt (W)	2.259 697	E-02
foot-pound per second (ft·lb/s)	watt (W)	1.355 818	
horsepower (hp)	watt (W)	7.456 999	E+02
horsepower [boiler]	watt (W)	9.809 500	E+03
horsepower [electric]	watt (W)	*7.460	E+02
horsepower [metric] (hp$_M$)	watt (W)	7.354 990	E+02
horsepower [water]	watt (W)	7.460 430	E+02
ton [refrigeration]	watt (W)	3.516 800	E+03

* Exact conversion.

Table 5-35.—Pressure—Stress (M/LT^2)

To convert from	To	Multiply by	
atmosphere [standard] (atm)	pascal (Pa)	1.013 250	E+05
bar	pascal (Pa)	*1.000	E+05
barye	pascal (Pa)	*1.000	E-01
dyne per square centimeter (dyn/cm^2)	pascal (Pa)	*1.000	E-01
foot of water [4 °C]	pascal (Pa)	2.988 980	E+03
gram force per square centimeter (gf/cm^2)	pascal (Pa)	*9.806 650	E+01
inch of mercury [0 °C]	pascal (Pa)	3.386 380	E+03
inch of mercury [16 °C]	pascal (Pa)	3.376 850	E+03
inch of water [4 °C]	pascal (Pa)	2.490 817	E+02
inch of water [16 °C]	pascal (Pa)	2.488 400	E+02
kilogram force per square meter (kgf/m^2)	pascal (Pa)	*9.806 650	
kilogram force per square centimeter (kgf/cm^2)	pascal (Pa)	*9.806 650	E+04

CHAPTER 5—UNIT AND FORMULAE CONVERSIONS

Table 5-35.—*Pressure—Stress* (M/LT^2)—Continued

To convert from	To	Multiply by
kip per square inch (kip/in^2)	pascal (Pa)	6.894 757 E+06
kip per square foot (kip/ft^2)	pascal (Pa)	4.788 026 E+04
megapascal (MPa)	pascal (Pa)	*1.000 E+06
meter-head [meter of water, 4 °C]	pascal (Pa)	9.806 365 E+03
millibar (mbar)	pascal (Pa)	*1.000 E+02
millimeter of mercury [0 °C] (mm(Hg))	pascal (Pa)	1.333 220 E+02
millimeter of water [4 °C] (mm(H$_2$O))	pascal (Pa)	9.806 365
newton per square meter (N/m^2)	pascal (Pa)	*1.000
pound per square foot (lb/ft^2)	pascal (Pa)	4.788 026 E+01
pound per square inch (lb/in^2)	pascal (Pa)	6.894 757 E+03
poundal per square foot (pdl/ft^2)	pascal (Pa)	1.488 164
tor	pascal (Pa)	*1.000
torr (mm(Hg))	pascal (Pa)	1.333 220 E+02

* Exact conversion.

Table 5-36.—Temperature (T)

Scale values	Degrees Celsius °C	Degrees Fahrenheit °F	Kelvins K	Degrees Rankine °R	Degrees Reaumur °r
$\chi\ °C =$	–	$\frac{9}{5}\chi + 32$	$\chi + 273.15$	$\frac{9}{5}\chi + 491.67$	$\frac{4}{5}\chi$
$\chi\ °F =$	$\frac{5}{9}(\chi - 32)$	–	$\frac{5}{9}(\chi + 459.67)$	$\chi + 459.67$	$\frac{4}{9}(\chi - 32)$
$\chi\ K =$	$\chi - 273.15$	$\frac{9}{5}\chi - 459.67$	–	$\frac{9}{5}\chi$	$\frac{4}{5}(\chi - 273.15)$
$\chi\ °R =$	$\frac{5}{9}(\chi - 491.67)$	$\chi - 459.67$	$\frac{5}{9}\chi$	–	$\frac{4}{9}(\chi - 491.67)$
$\chi\ °r =$	$\frac{5}{4}\chi$	$\frac{9}{4}\chi + 32$	$\frac{5}{4}\chi + 273.15$	$\frac{9}{4}\chi + 491.67$	–

Intervals:

	°C	K	°F	°R	°r
$1\ °C = 1\ K =$	1		$\frac{9}{5}$		$\frac{4}{5}$
$1\ °F = 1\ °R =$	$\frac{5}{9}$		1		$\frac{4}{9}$
$1\ °r =$	$\frac{5}{4}$		$\frac{9}{4}$		1

CHAPTER 5—UNIT AND FORMULAE CONVERSIONS   217

Table 5-37.—*Time* (*T*)

To convert from	To	Multiply by	
day [mean solar] (d)	second (s)	*8.640	E+04
day [sidereal]	second (s)	8.616 409	E+04
hour [mean solar] (hr)	second (s)	*3.600	E+03
hour [sidereal]	second (s)	3.590 170	E+03
minute [mean solar] (min)	second (s)	*6.000	E+01
minute [sidereal]	second (s)	5.983 617	E+01
month [mean calendar] (mo)	second (s)	*2.628	E+06
second [sidereal]	second (s)	9.972 696	E−01
week [7 days] (wk)	second (s)	*6.048	E+05
year [calendar] (a)	second (s)	*3.153 600	E+07
year [sidereal]	second (s)	3.155 815	E+07

* Exact conversion.

Table 5-38.—*Torque—Bending moment* (ML^2/T^2)

To convert from	To	Multiply by	
dyne centimeter (dyn·cm)	newton meter (N·m)	*1.000	E−07
kilogram force meter (kgf·m)	newton meter (N·m)	*9.806 650	
kip-foot (kip·ft)	newton meter (N·m)	1.355 818	E+02
ounce inch (oz·in)[a,b]	newton meter (N·m)	7.061 552	E−03
pound-foot (lb·ft)[a,b]	newton meter (N·m)	1.355 818	
pound-inch (lb·in)[a,b]	newton meter (N·m)	1.129 848	E−01

* Exact conversion.
[a] The addition of the force designator may be desirable, e.g., lbf·ft.
[b] Most USBR engineers reverse the torque units, for example, foot-pound; this is equivalent terminology.

Table 5-39.—*Transmissivity* (L^3/LT)

To convert from	To	Multiply by	
cubic foot per foot day[a] ($ft^3/(ft \cdot d)$)	cubic meter per meter year[b] ($m^3/(m \cdot a)$)	3.390 861	E+01
cubic meter per meter day[c] ($m^3/(m \cdot d)$)	cubic meter per meter year ($m^3/(m \cdot a)$)	3.650	E+02
darcy-foot	cubic meter per meter year ($m^3/m \cdot a$)	8.871	E+02
gallon per foot day ($gal/(ft \cdot d)$)	cubic meter per meter year ($m^3/(m \cdot a)$)	4.533 057	
liter per meter day ($L/(m \cdot d)$)	cubic meter per meter year ($m^3/(m \cdot a)$)	3.650	E-01

[a] Also written as square foot per day (ft^2/d).
[b] Also written as square meter per year (m^2/a).
[c] Also written as square meter per day (m^2/d).

Table 5-40.—*Velocity—Speed*

To convert from	To	Multiply by	
Angular (θ/T):			
degree per second	radian per second (rad/s)	1.745 329	E-02
revolution per minute (r/min)	radian per second (rad/s)	1.047 198	E-01
revolution per second (r/s)	radian per second (rad/s)	6.283 185	
Linear (L/T):			
foot per second (ft/s)	meter per second (m/s)	*3.048	E-01
foot per minute (ft/min)	meter per second (m/s)	*5.080	E-03
foot per hour (ft/h)	meter per second (m/s)	8.466 667	E-05
foot per day (ft/d)	meter per second (m/s)	3.527 778	E-06
foot per year (ft/a)	meter per second (m/s)	9.695 890	E-09
inch per second (in/s)	meter per second (m/s)	*2.540	E-02
inch per hour (in/h)	meter per second (m/s)	7.055 556	E-06
kilometer per hour (km/h)	meter per second (m/s)	2.777 778	E-01
knot [nautical miles per hour] (kn)	meter per second (m/s)	5.144 444	E-01
mile per hour (mi/h)	meter per second (m/s)	4.470 400	E-01
meter per hour (m/h)	meter per second (m/s)	2.777 778	E-04
meter per year (m/a)	meter per second (m/s)	3.170 979	E-08

CHAPTER 5—UNIT AND FORMULAE CONVERSIONS 219

Table 5-40.—*Velocity—Speed*—Continued

To convert from	To	Multiply by	
millimeter per second (mm/s)	meter per second (m/s)	*1.000	E-03
speed of light (c)	meter per second (m/s)	2.997 925	E+08

* Exact conversion.

Table 5-41.—*Viscosity*

To convert from	To	Multiply by	
Kinematic (L^2/T):			
centistoke (cSt)	square meter per second (m^2/s)	*1.000	E-06
square foot per second (ft^2/s)	square meter per second (m^2/s)	9.290 304	E-02
stokes (St)	square meter per second (m^2/s)	*1.000	E-04
Dynamic (M/LT):			
centipoise (cP)	pascal second (Pa·s)	*1.000	E-03
poise (P)	pascal second (Pa·s)	*1.000	E-01
poundal second per square foot (pdl·s/ft^2)	pascal second (Pa·s)	1.488 164	
pound-force second per square foot (lbf·s/ft^2)	pascal second (Pa·s)	4.788 026	E+01
pound per foot-hour (lb/(ft·h))	pascal second (Pa·s)	4.133 789	E-04
pound per foot-second (lb/(ft·s))	pascal second (Pa·s)	1.488 164	
reciprocal rhe (rhe^{-1})	pascal second (Pa·s)	*1.000	E-01
slug per foot-second (slug/(ft·s))	pascal second (Pa·s)	4.788 026	E+01

* Exact conversion.

Table 5-42.—Volume—Capacity (L^3)

To convert from	To	Multiply by	
acre-foot [U.S. survey] (acre-ft)	cubic meter (m^3)	1.233 489	E+03
barrel [oil] (bbl)	cubic meter (m^3)	1.589 873	E-01
barrel [water] (bbl)	cubic meter (m^3)	1.192 405	E-01
board foot [1 ft x 1 ft x 1 in] (bm)	cubic meter (m^3)	2.359 737	E-03
bushel [U.S., dry] (bu)	cubic meter (m^3)	3.523 907	E-02
cord	cubic meter (m^3)	3.624 556	
cubic centimeter (cm^3)	cubic meter (m^3)	*1.000	E-06
cubic decimeter (dm^3)	cubic meter (m^3)	*1.000	E-03
cubic dekameter (dam^3)	cubic meter (m^3)	*1.000	E+03
cubic foot (ft^3)	cubic meter (m^3)	2.831 685	E-02
cubic inch (in^3)	cubic meter (m^3)	1.638 706	E-05
cubic kilometer (km^3)	cubic meter (m^3)	*1.000	E+09
cubic mile (mi^3)	cubic meter (m^3)	4.168 182	E+09
cubic millimeter (mm^3)	cubic meter (m^3)	*1.000	E-09
cubic yard (yd^3)	cubic meter (m^3)	7.645 549	E-01
cup	cubic meter (m^3)	2.359 737	E-03
firkin	cubic meter (m^3)	3.406 871	E-02
fluid dram	cubic meter (m^3)	3.696 691	E-06
fluid ounce [U.S.] (fl.oz.)	cubic meter (m^3)	2.957 353	E-05
gallon [Imperial]	cubic meter (m^3)	4.546 060	E-03
gallon [U.S., dry]	cubic meter (m^3)	4.404 884	E-03
gallon [U.S., liquid] (gal)	cubic meter (m^3)	3.785 412	E-03
gill [U.S.]	cubic meter (m^3)	1.182 941	E-04
kiloliter (kL)	cubic meter (m^3)	*1.000	
liter (L)	cubic meter (m^3)	*1.000	E-03
megaliter (ML)	cubic meter (m^3)	*1.000	E+03
milliliter (mL)	cubic meter (m^3)	*1.000	E-06
peck [U.S.]	cubic meter (m^3)	8.809 768	E-03
pint [U.S., dry]	cubic meter (m^3)	5.506 105	E-04
pint [U.S., liquid]	cubic meter (m^3)	4.731 765	E-04
quart [U.S., dry]	cubic meter (m^3)	1.101 221	E-03
quart [U.S., liquid]	cubic meter (m^3)	9.463 529	E-04
stere [timber]	cubic meter (m^3)	*1.000	
tablespoon	cubic meter (m^3)	1.478 676	E-05
teaspoon	cubic meter (m^3)	4.928 922	E-06
ton [sea freight or shipping capacity]	cubic meter (m^3)	1.132 674	
ton [internal cap. of ships or register ton]	cubic meter (m^3)	2.831 685	
ton [vol. of oil]	cubic meter (m^3)	6.700 179	
ton [timber]	cubic meter (m^3)	1.415 842	
tun [U.S., liquid]	cubic meter (m^3)	9.539 238	E-01

* Exact conversion.

Chapter VI

ENGINEERING PROBLEMS AND FORMULAE

Introduction

Example problems and formulae using SI units are presented to acquaint Reclamation technical personnel with typical situations. The problems/formulae selected are those considered most representative of the type of work performed by the Bureau. Not all problem types or fields of engineering are included; this is due to a desire not to duplicate available engineering texts.

For additional problems, the interested individual should review the ASME text booklet series [4 through 10], or the British Ministry of Public Works' "Metrication in the Construction Industry, No. 2" [11].

Mechanics and Dynamics

Mass and Force Problems

Note: $F = m \times a$

Standard earth gravitational acceleration $(g) = 9.806\ 65$ m/s^2; use 9.81 m/s^2 for engineering calculations.

(a) What mass exerts a gravitational force of 78.5 N at earth mean sea level?

$F = m \times a$
78.5 N = $m \times$ 9.81 m/s^2

$$m = \frac{78.5}{9.81} = 8 \text{ kg}$$

(b) What is the acceleration imparted to a mass of 200 g by a force of 1.5 kN?

200 g = 0.2 kg
1.5 kN = 1500 N
$F = m \times a$
1500 N = 0.2 kg $\times a$

$$a = \frac{1500}{0.2} = 7500 \text{ m/s}^2$$

(c) What is the gravitational force of a 10-kg bag of seed?

$F = m \times a$
$F = 10 \text{ kg} \times 9.806\ 65 \text{ m/s}^2$
$= 98 \text{ N}$

(d) A floating log with a mass of 2800 kg strikes a spillway gate with considerable velocity and is brought to a halt with a uniform deceleration of 180 m/s². If the gate supports are commercially rated to withstand a force of 95 000 pounds, will the gate supports in fact be damaged?

$F = m \times a$
$= 2800 \times 180$
$= 504 \text{ kN}$

To convert pounds force to kilonewtons, multiply by 4.45×10^{-3}
Therefore, allowable force = $95\ 000 \times 4.45 \times 10^{-3} = 422.8$ kN
Thus, the gate will be damaged.

(e) Water is flowing through a pipe. The force exerted on a 90° bend in the direction of the inflowing water due to the change in momentum of the water passing around the bend is:

$F = QwV$

where:

Q = flow rate (m³/s)
w = water density (kg/m³)
V = flow velocity (m/s)

Derivation:

Force = change of momentum in unit time

Change in momentum of unit mass = $1 (V - 0)$
= $1V$
Mass passing around bend per second = $(Q) (w)$
∴ Change of momentum per second = $(Q) (w) (1V)$
= QwV

Calculate the force for the following data:

$Q = 1.5$ m³/s
$w = 1000$ kg/m³
$V = 2$ m/s

CHAPTER 6—ENGINEERING PROBLEMS AND FORMULAE

Solution:

$F = QwV$
$= 1.5 \times 1000 \times 2$
$= 3000$ newtons

The preceding examples illustrate that in the computation of dynamic forces on structures, the procedure is shorter [i.e., dividing by the gravitational acceleration (g) is eliminated] if absolute units of force (newton) are used instead of gravitational force units.

Active Cohesionless Soil Pressure

$$P = \left(\frac{1 - \sin \theta}{1 + \sin \theta}\right) wZ$$

where:

P = active pressure for a cohesionless soil
θ = angle of internal friction
w = soil density
Z = depth of soil to point considered

In customary units, P is in pounds per square foot (lb/ft^2), w is in pounds per cubic foot (lb/ft^3), and Z is in feet. Dimensionally, the equation can be written as:

$$\frac{lb}{ft^2} = \frac{lb}{ft^3} \times ft$$

This appears to balance, but if dimensional symbols were used instead of unit symbols, this equation would look like this:

$$\begin{array}{ccccccc}
 & & (\text{density}) & & (\text{length}) & & (\text{adjustment factor}) \\
\dfrac{M \times L/T^2}{L^2} & = & \dfrac{M}{L^3} & \times & L & \times & K \\
\dfrac{M}{LT^2} & = & \dfrac{M}{L^3} & \times & K & & \\
K & = & L/T^2 & & & &
\end{array}$$

The adjustment factor, K, has the dimensions of acceleration. The reason this shows up in the dimensional analysis and not in the unit analysis is the dual definition of the pound. Note that soil density is a mass per volume unit, not a force per volume. To adjust this formula for SI units, the density must be multiplied by gravitational acceleration.

$$P = \left(\frac{1 - \sin\theta}{1 + \sin\theta}\right) 9.81\, wZ$$

where:

P = pressure, in pascals
θ = plane angle, in degrees or radians
w = soil density, in kilograms per cubic meter
Z = soil depth, in meters
9.81 = gravitational constant, in meters per second squared

Drop Hammer Loading

$$Q = \frac{1634\, WH}{s + c}$$

where:

Q = safe design load on a pipe, in newtons
W = mass of the driver ram, in kilograms
H = fall height of the ram, in meters
s = the pile penetration under the last blow, in millimeters
c = a constant equal to 25.4 for a drophammer

Note that the gravitational constant has been factored into the multiplication coefficient (1634); the coefficient must be changed if the gravitational acceleration differs significantly from 9.81 m/s^2.

$$Q = \frac{1634\, WH}{(s + 25.4)}$$

Pump Power

$$P_o = Q\rho g H$$

where:

P_o = pump power, in watts
Q = flow rate, in cubic meters per second
H = head, in meters
ρ = density, in kilograms per cubic meter
g = gravitational acceleration, in meters per second squared

CHAPTER 6—ENGINEERING PROBLEMS AND FORMULAE

$$P_o(W) = P_o \left(\frac{kg \cdot m^2}{s^3}\right)$$

$$= Q\left(\frac{m^3}{s}\right) \times H(m) \times \rho\left(\frac{kg}{m^3}\right) \times g\left(\frac{m}{s^2}\right)$$

Vibration

(a) Natural frequency

$$\omega_n = \sqrt{\frac{gk}{W}}$$

where:

ω_n = natural frequency, in hertz
k = stiffness constant, in newtons per meter
g = gravitational acceleration, in meters per second squared
W = gravitational force exerted by the machine, in newtons

Note that W/g can be simplified to a mass term expressed in kilograms.

$$\omega_n = \sqrt{\frac{k}{M}}$$

(b) Amplitude of vibration

$$X = \frac{10^3 \, F_o \, k^{-1}}{\left[\left[1 - \left(\frac{\omega}{\omega_n}\right)^2\right]^2 + \left(2\delta \frac{\omega}{\omega_n}\right)^2\right]^{0.5}}$$

where:

X = vibration amplitude (mm)
F_o = magnitude of cyclic force (N)
k = spring constant (N/m)
δ = damping factor (dimensionless)
$\frac{\omega}{\omega_n}$ = ratio of cyclic force to natural frequency (dimensionless)

(c) Transmissibility ratio

$$T.R. = \frac{\left[1 + \left(2\delta \frac{\omega}{\omega_n}\right)^2\right]^{0.5}}{\left[\left[1 - \left(\frac{\omega}{\omega_n}\right)^2\right]^2 + \left(2\delta \frac{\omega}{\omega_n}\right)^2\right]^{0.5}}$$

Structures

Beam Deflection

$$d = \frac{PL^3}{48\,EI}$$

where:

d = deflection at midspan, in meters
P = concentrated load of center of span, in newtons
L = total span, in meters, of simply supported beam
E = modulus of elasticity, in pascals
I = moment of inertia, in m^4
48 = a constant

It was not necessary to change this equation when converting from the customary units, because base units were substituted.

If the following units were used,

d — millimeters
P — kilonewtons
L — meters
E — gigapascals
I — mm^4

the equation is:

$$d = \frac{PL^3}{48\,EI} \times 10^9$$

Beam Shear Stress

$$v = \frac{V}{b\,m_s}$$

where:

v = shear stress in a beam, in pascals
V = shear force, in newtons
b = breadth of beam, in meters
m_s = length of section moment arm, in meters

Pipe (Beam) Stresses

A metal pipe spanning a distance of 7 m and carrying water at low pressure is subjected to a maximum bending moment of 28 kN·m due to self load plus water load (mass forces).
The outside and inside diameters of the pipe are 650 and 635 mm, respectively. (As previously explained, it is preferred that all calculations be done in base units; therefore, these values will be converted to meters.)
(a) What is the maximum stress in the pipe in megapascals?
(b) The pipe is commercially rated at 20 000 lb/in^2. Convert this to megapascals and comment on the actual level of stress in the pipe.

Solution:

(a)
$$f = M/Z$$

where:

f = stress, in pascals (Pa)
M = bending moment, in newton meters (N·m)
Z = section modulus, in meters cubed (m^3)

$$Z = \frac{2 \times I_{cc}}{D}$$

where:

I_{cc} = moment of inertia about neutral axis (m^4)

$$I_{cc} = \frac{\pi}{64}(D^4 - d^4)$$

where:

D = outside diameter of pipe, in meters
d = inside diameter of pipe, in meters
$D^4 = (0.650)^4 = 0.178\ 506$
$d^4 = (0.635)^4 = 0.162\ 590$

$$I_{cc} = \frac{\pi}{64}(D^4 - d^4)$$

$$= \frac{\pi}{64}(0.178\ 506 - 0.162\ 590) = \frac{\pi}{64}(159.16 \times 10^{-4})$$

$$= 7.81 \times 10^{-4}\ m^4$$

$$Z = \frac{2 \times I_{cc}}{D}$$

$$= \frac{2 \times 7.81 \times 10^{-4}}{0.650}$$

$$= 2.40 \times 10^{-3} \text{ m}^3$$

$$f = \frac{M}{Z}$$

$$= \frac{28\,000}{2.40 \times 10^{-3}}$$

$$= 11\,666\,667 \text{ Pa}$$
$$= 11.67 \text{ MPa}$$

(b) To convert from pounds per square inch to megapascals, multiply by 6.895×10^{-3}.

$$20\,000 \text{ lb/in}^2 = 138 \text{ MPa}$$

Actual stress in the pipe wall is comparatively low.

Bending Moment

A beam, simply supported over a span of 8.5 m, is to support a mass loading of 260 kg/m distributed evenly along the beam.
(a) What is the force of gravity imposed on the beam in newtons per meter?
(b) What is the maximum bending moment on the beam in newton meters and kilonewton meters?

$$\text{Use } M = \frac{w \times L^2}{8}$$

where:

M = bending moment, in newton meters (N·m)
w = uniform load, in newtons per meter (N/m)
L = span, in meters (m)

Solutions:

(a) Uniform load $= 260 \text{ kg/m} \times 9.81 \text{ m/s}^2$ ($F = m \times a$ and $a = g = 9.81 \text{ m/s}^2$)
$= 2550.6 \text{ N/m}$

CHAPTER 6—ENGINEERING PROBLEMS AND FORMULAE

(b)
$$M = \frac{w \times L^2}{8}$$

$$= \frac{2550.6 \times (8.5)^2}{8}$$

$$= 23\,035 \text{ N·m}$$

Moment = 23.0 kN·m

Concrete Beam

A concrete beam resists an axial compressive force (F_c) of 50 kN and a bending moment (M) of 38 kN·m. The dimensions of the beam are given below and illustrated in figure 6-1.

b (width)   = 150 mm = 0.15 m
h (depth)   = 300 mm = 0.3 m
d (depth of tensile = 260 mm = 0.26 m
reinforcement)

(a) Determine the moment about the tensile reinforcement.

$$M' = M + F_c \left(d - \frac{h}{2} \right)$$

$$= (38 \times 10^3) + (50 \times 10^3)\left(0.26 - \frac{0.3}{2}\right)$$

$$= (38 \times 10^3) + (50 \times 10^3)(0.11)$$

$$= 43.5 \times 10^3 \text{ N·m}$$

$$= 43.5 \text{ kN·m}$$

Figure 6-1.—Concrete beam in axial compression with a bending moment. 40-D-6361

(b) Determine if compressive reinforcement is required.

$$M_c = Q_c bd^2$$

where:

M_c = maximum bending moment which beam can safely resist without compression reinforcement
Q_c = coefficient for resistance moment based on concrete strength

b & d as previously defined

$$Q_c = \left(\frac{n_1 a_1}{2}\right) f_{cb}$$

where:

n_1 = depth factor of neutral axis
a_1 = lever arm factor
f_{cb} = allowable compressive stress in concrete due to bending

$$n_1 = \frac{1}{\left(1 + \dfrac{f_{st}}{a_e f_{cb}}\right)} \qquad a_1 = \left(1 - \frac{n_1}{3}\right)$$

where:

f_{st} = allowable tensile stress in steel
a_e = ratio of Young's modulus, steel:concrete

From engineering reference materials:

$f_{st} = 140 \times 10^6$ Pa (140 MPa)
$f_{cb} = 8 \times 10^6$ Pa   (8 MPa)
$a_e = 15$

Substituting:

$$n_1 = \frac{1}{1 + \dfrac{140(10^6)}{15(8)(10^6)}} = 0.4615$$

$$a_1 = 1 - (0.4615/3) = 0.8462$$

$$Q_c = \frac{(0.4615)(0.8462)(8 \times 10^6)}{2}$$

$$= 1.562 \times 10^6 \text{ Pa}$$

CHAPTER 6—ENGINEERING PROBLEMS AND FORMULAE

$$M_c = (1.562 \times 10^6)(0.15)(0.26)^2$$
$$= 15.84 \times 10^3 \text{ N·m} = 15.84 \text{ kN·m}$$

As this is less than M', compressive reinforcement will be required.

(c) Determine what tensile reinforcement will be required.

$$A_s = \left(\frac{M'}{f_{st} \, a_1 \, d}\right) - \left(\frac{F_c}{f_{st}}\right)$$

where:

A_s = area of tensile reinforcement

$$A_s = \left(\frac{(43.5 \times 10^3)}{(140 \times 10^6)(0.8462)(0.26)}\right) - \left(\frac{50 \times 10^3}{140 \times 10^6}\right)$$
$$= 0.1055 \times 10^{-2} \text{ m}^2$$
$$= 1055 \text{ mm}^2$$

Using three bars as illustrated (fig. 6-1) requires that each has a minimum diameter of 22 mm to adequately reinforce the concrete.

Prestressed Concrete

Symbols used:

A	= cross-sectional area
M	= bending moment
P	= prestressing force in tendon
Z	= section modulus
L	= span of beam
e	= eccentricity of prestressing cable
f	= allowable tensile or compressive stress
$1 - B$	= portion of prestressing force attributable to all losses
σ	= calculated tensile or compressive stress
ν	= allowable live load

Problem:
Determine the allowable live load on a prestressed concrete beam. The beam is simply supported on a span of 16.3 m. The initial tendon force is 1.78 MN with an eccentricity of 380 mm. After all losses, 90 percent of the initial prestressing force is retained. The concrete has a mass density of 2380 kg/m³. The allowable stresses in concrete are 13.85 MPa in compression and 1.88 MPa in tension.

For the section shown in figure 6-2, the following properties can be calculated:

Sectional area $A = 0.335 \text{ m}^2$
Section modulus $Z = 82\,080 \times 10^{-6} \text{ m}^3$

The force per unit length of the beam is:

density $\times A \times g$
$(2380)\,(0.335)\,(9.81) = 7.819 \text{ kN/m}$

Dead load bending moment equals:

force per unit length $\times \dfrac{L^2}{8}$

$(7.819 \times 10^3) \times \dfrac{(16.3)^2}{8} = 260 \text{ kN·m}$

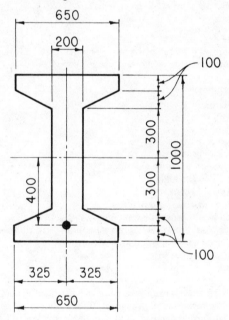

Dimensions are in millimeters unless otherwise shown.

Figure 6-2.—Concrete I-beam with reinforcing tendon. 40-D-6362

CHAPTER 6—ENGINEERING PROBLEMS AND FORMULAE   233

Top and bottom fiber stresses due to dead load:

$$\pm \frac{M}{Z} = \pm \frac{260 \times 10^3}{82\,080 \times 10^{-6}} \qquad \begin{pmatrix} + \text{ means compression} \\ - \text{ means tension} \end{pmatrix}$$

$$= \pm 3.16 \text{ MPa}$$

Tendon force before losses:

$$P_o = 1.78 \times 10^6 \text{ N}$$

Tendon force after losses:

$$P = P_o \times \text{retained prestressing percentage} = (1.78 \times 10^6)(0.90) = 1.6 \text{ MN}$$

Top and bottom fiber stresses due to tendon force (after losses):

$$\sigma = \frac{P}{A} \pm \frac{Pe}{Z} = \frac{1.6 \times 10^6}{0.335} \pm \frac{(1.6 \times 10^6)(380 \times 10^{-3})}{(82\,080 \times 10^{-6})}$$

$$= (4.78 \times 10^6) \pm (7.41 \times 10^6)$$

$$= 4.78 \pm 7.41 \text{ MPa}$$

Top: -2.63 MPa
Bottom: $+12.19$ MPa

Dead load plus tendon stresses:

Top: $-2.63 + 3.16 = 0.53$ MPa
Bottom: $+12.19 - 3.16 = 9.03$ MPa

For live load, the available change is the lesser of the following:

Top: $13.85 - 0.53 = 13.32$ MPa   ($f_{top} = 13.85$ MPa)
Bottom: $1.88 + 9.03 = 10.91$ MPa   ($f_{bottom} = 1.88$ MPa)

The allowable live load bending moment is:

$$M = \sigma Z$$
$$= (10.91 \times 10^6)(82\,080) \times 10^{-6} = 895\,493 \text{ N·m}$$
$$= 895 \text{ kN·m}$$

Therefore, the allowable live load is:

$$v = \frac{8M}{L^2}$$

$$= \frac{8(895 \times 10^3)}{(16.3)^2} = 26\,949 \text{ N/m}$$

$$= 26.95 \text{ kN/m}$$

Hydrology

Fluid Dynamics

(a) Static fluid pressure
A tank contains water 12.5 meters deep. The tank has a flat bottom surface area of 3.7 m².
 (i) What is the mass of the water inside the tank?
 (ii) What force does this mass of water exert on the tank floor?
 (iii) What is the pressure being exerted on the bottom surface?

Notes: Density of water = 1000 kg/m³
 (i) The volume of the water = 12.5 m × 3.7 m² = 46.25 m³

 The mass of the volume of water = 1000 kg/m³ × 46.25 m³
 = 46 250 kg

 (ii) Newton's law states $F = m \times a$

 $a = g = 9.81$ m/s²
 Force = 46 250 kg × 9.81 m/s²
 = 453 712 newtons

 (iii) Pressure = force per unit area

 = 453 712 N ÷ 3.7 m²
 = 122 625 pascals
 ≈ 122.6 kPa

(b) Bernoulli's equation
A simple statement of Bernoulli's equation is "in a flowing liquid, provided that there are no energy inputs or outputs between two cross sections and friction is negligible, the sum of the velocity head, the datum head, and the pressure head is the same at both sections."

CHAPTER 6—ENGINEERING PROBLEMS AND FORMULAE

Velocity head = $\dfrac{V^2}{2g}$

Datum head = Z

Pressure head = P/γ (γ = specific force)

$$\dfrac{V_1^2}{2g} + Z_1 + \dfrac{P_1}{\gamma} = \dfrac{V_2^2}{2g} + Z_2 + \dfrac{P_2}{\gamma}$$

Problem:
The flow in a pipe is 0.61 m^3/s. At section 1, the velocity is negligible, the height of the pipe centerline is 46 m above a datum level, and the gage pressure in the pipe is 136 kPa. At section 2, the cross sectional area of the pipe is 0.04 m^2 and the height of the pipe centerline is 51 m above the datum. What would be the gage pressure reading in kilopascals at section 2? (A reading of 0 kPa indicates atmospheric pressure, a positive reading indicates pressure greater than atmospheric, and a negative reading denotes a subatmospheric pressure in which case a vacuum gage would be needed.)

Section 1:

Velocity head (V_1) = 0 m

Datum head (Z_1) = 46 m

Pressure head $\left(\dfrac{P_1}{\gamma}\right) = \dfrac{136}{9.79} = 13.89$ m   (γ = 9.79 kN/m^3 @ 20 °C)

Section 2:

Flow = 0.61 m^3/s   Velocity = $\dfrac{\text{Flow}}{\text{Area}} = \dfrac{0.61}{0.04} = 15.25$ m/s

Velocity head (V_2) = $\dfrac{(15.25)^2}{2(9.81)} = 11.85$ m

Datum head (Z_2) = 51 m

Pressure head (P_2/γ) = ? m

Bernoulli's equation:

$$0 + 46 + 13.89 = 11.85 + 51 + \dfrac{P_2}{(9.79)}$$

$$\frac{P_2}{(9.79)} = 59.89 - 62.85 = -2.96$$

$P_2 = -29.0\,\text{kPa (gage)}$
$\quad\ = 72.0\,\text{kPa (absolute)}$

The absolute pressure is high enough to prevent the gases coming out of solution and causing vapor locks.

(c) Pipe constriction head loss

$$H = K\,\frac{V^2}{2g}$$

where:

H = head loss at sudden pipe constriction
V = velocity
g = standard gravitational acceleration
K = coefficient whose value depends on ratio of D_2/D_1

Dimensional representation:

$$L = \left(\frac{L^2}{T^2}\right) \times (L/T^2)^{-1}$$

$$L = L$$

The coefficient is dimensionless; the formula may be used as written with velocity in meters per second (m/s) and gravitational acceleration in meters per second squared (m/s^2). The head will be in meters; to convert to kilopascals, multiply by 9.81 and the relative mass density.

(d) Manning's equation

$$Q = \frac{1}{n} A r^{0.667} S^{0.5}$$

where:

Q = discharge (m^3/s)
n = roughness coefficient
A = cross-sectional area (m^2)
r = hydraulic radius (m), cross-sectional area divided by the wetted perimeter
S = slope of channel (dimensionless decimal number)

CHAPTER 6—ENGINEERING PROBLEMS AND FORMULAE   237

Problem:
A concrete pipe ($n = 0.017$), flowing fully, has a diameter of 800 mm and a hydraulic gradient of 1.25×10^{-3}. Using Manning's equation:
(i) Calculate the flow velocity in meters per second.
(ii) What is the flow rate in cubic dekameters per day?

Solution:

(i) $$V = \frac{r^{0.667} \sqrt{S}}{n}$$

where:

V = flow velocity (m/s)
r = hydraulic radius (m)
S = hydraulic gradient (dimensionless)
n = coefficient
Hydraulic radius is one-fourth of the pipe diameter.

$$r = \frac{800}{4} = 200 \text{ mm} = 0.2 \text{ m}$$

knowns:

$S = 1.25 \times 10^{-3}$
$n = 0.017$

$$V = \frac{(0.2)^{0.667} \sqrt{1.25 \times 10^{-3}}}{0.017}$$

$$= \frac{(0.342)(35.35 \times 10^{-3})}{0.017} = 0.71 \text{ m/s}$$

(ii) $\qquad\qquad\qquad Q = AV$

where:

A = pipe area (m^2)
V = flow velocity (m/s)
$A = \pi d^2/4 = \pi (0.8)^2/4$
$\quad = 0.503$ m^2
$Q = 0.503 \times 0.71$
$\quad = 0.357$ m^3/s
To convert from cubic meters per second (m^3/s) to cubic dekameters per day (dam^3/d), multiply by 86.4.

$$Q = 0.357 \times 86.4 = 30.8 \text{ dam}^3$$

Refer to figure 6-3, this is an engineering nomograph of Manning's equation; it can be used to solve a reasonably broad range of problems with about a 98-percent accuracy level. Flow rate can be determined by multiplying the cross-sectional area times the calculated flow velocity.

Flood Hydrology

(a) Rational formula—Used to estimate peak discharge for small drainages under certain circumstances

$$Q_p = 2.755 \times 10^{-3} \, CIA$$

where:

Q_p = peak discharge (m³/s)
C = basin characteristics coefficient, value ranges between 0 and 1.0
I = rainfall intensity (mm/h), average for a period of time equal to the time of concentration
A = drainage area (ha)

(b) Peak discharge in a triangular hydrograph

$$Q_p = 2.083 \times 10^{-3} \, AQ/T_p$$

where:

Q_p = peak discharge (m³/s)
A = drainage area (ha)
T_p = time length of hydrograph base (h)
Q = runoff depth (mm) per unit area

(c) Concentration time

$$T_c = 0.948 \left(\frac{L^3}{h}\right)^{0.385}$$

where:

T_c = time of concentration in hours (h)
L = length of longest watercourse (km)
h = elevation differential (m)

(d) Lag time

$$T_L = C \left(\frac{31.6 \, LL_{ca}}{\sqrt{S}}\right)^k$$

CHAPTER 6—ENGINEERING PROBLEMS AND FORMULAE

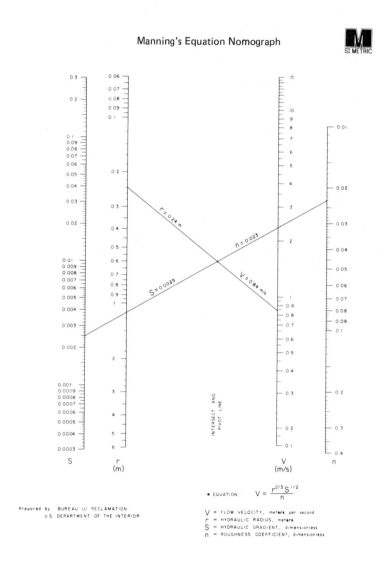

Figure 6-3.—Nomograph for Manning's equation. 40-D-6363

where:

T_L = lag time (h)
C and k = coefficients characteristic of the drainage basin type
L = length of longest watercourse (km)
S = average slope of watercourse (dimensionless decimal number)
L_{ca} = length of watercourse to area centroid (km)

(e) Myer rating

$$q = C\sqrt{A}$$

where:

q = floor peak in $(ft^3/s)/mi^2$ $[(m^3/s)/km^2]$
C = Myer coefficient
A = catchment area in square miles (square kilometers)

For this problem assume that the customary value of C for a particular catchment is 30. What is it for SI units?

Dimensionally:

$$\frac{L^3}{TL^2} = C\sqrt{L^2}$$

$$C = T^{-1}$$

Also, nonbasic units are used; conversion will be necessary.

$$(ft^3/s)/mi^2 = 30\sqrt{mi^2} \qquad (1)$$

$$(m^3/s)/km^2 = C_{SI}\sqrt{km^2} \qquad (2)$$

Divide (1) by (2).

$$\left(\frac{ft^3}{m^3}\right)\left(\frac{km^2}{mi^2}\right) = \frac{30}{C_{SI}}\sqrt{\frac{mi^2}{km^2}}$$

$$C_{SI} = 30\left(\frac{mi}{km}\right)^3\left(\frac{m}{ft}\right)^3$$

km = 1.6093 × mi   (mile to kilometer conversion)
m = 0.3048 × ft   (feet to meter conversion)

CHAPTER 6—ENGINEERING PROBLEMS AND FORMULAE   241

Substituting

$$C_{SI} = 30 \left(\frac{mi}{1.6093 \times mi}\right)^3 \left(\frac{0.3048 \times ft}{ft}\right)^3$$

$$= 30(0.6214)^3 \ (0.3048)^3$$

$$= 30(0.24) \ (0.0283)$$

$$= 0.204$$

$$q = 0.204 \sqrt{A} \qquad \text{(SI units)}$$

Sedimentation

(a) DuBoy's formula

$$Q_{BM} = 119\ 054\ D^{-0.75}\ (3.2808\ dS)^2\ w$$

where:

Q_{BM} = discharge of bed material, measured in metric tons per day (t/d)
D = median diameter of bed material, in millimeters (mm)
d = mean flow depth, in meters (m)
S = slope of the energy gradient, dimensionless
w = flow width, in meters (m)

(b) Schoklitsch formula

$$Q_{BM} = 20\ 814\ D^{-0.5}\ S^{1.5} \left(35.31\ Q - (6.87 \times 10^{-4}\ wD)\ (S^{-1.333})\right)$$

where:

Q = liquid discharge, in cubic meters per second (m³/s); all other quantities are as defined in DuBoy's formula (a).

(c) Tractive force equation

$$TF = 9.81\ \gamma dS$$

where:

TF = tractive force (N/m²)
γ = unit mass density of liquid (kg/m³)
d = mean flow depth, in meters (m)
S = slope of energy gradient, dimensionless

(d) Tractive power equation

$$TP = (TF)V$$

where

TP = tractive power [W/m^2 or (N/m)/s]
V = mean velocity of flow (m/s)
TF = tractive force (N/m^2)

(e) Meyer-Peter Muller equation

$$\gamma_w \left(\frac{Q_s}{Q}\right)\left(\frac{k_s}{k_r}\right)^{1.5} dS = 0.047\, \gamma_s'\, D_m + 0.25 \sqrt[3]{\frac{\gamma_w g_s'^2}{g}}$$

where:

γ_w = specific mass of water (kg/m^3)
Q_s = water discharge rate determining the bed load transport (m^3/s)
Q = total water discharge rate (m^3/s)
k_s = the bed roughness parameter ($\sqrt[3]{m}$/s)
k_r = particle roughness, $C/\sqrt[6]{D_{90}}$, ($\sqrt[3]{m}$/s) where C = 26, ($\sqrt{m}$/s)
d = depth of flow, in meters (m)
S = slope of the energy gradient, dimensionless
γ_s = specific mass of sediment (out of water)
$\gamma_s' = \gamma_s - \gamma_w$, specific mass of sediment in water (kg/m^3)
D_m = effective diameter of bed material (m), equal to $\frac{\Sigma D \Delta p}{100}$
where D is average size of particles in a size fraction and p is the percent in that size fraction
g = gravitational acceleration (m/s^2)
g_s' = the specific bedload transport mass under water [(kg/s)/m]; the meter dimension refers to the stream width

Water Utilization

(a) Jensen-Haise equation

$$ETP = 0.035\, C_T(T_o - T_x)\, R_s$$

where:

ETP = potential evapotranspiration (mm/d)
C_T = temperature coefficient
T_x = intercept of the temperature axis

CHAPTER 6—ENGINEERING PROBLEMS AND FORMULAE

T_o = mean air temperature (°C)
R_s = mean daily observed shortwave radiation rate (W/m²)
Note: Presently, the observed shortwave radiation is recorded in langleys per day. For these dimensioned data, the equation becomes

$$ETP = 0.0171\ C_T (T_o - T_x)\ R_s$$

(b) Penman equation

$$ETP = \frac{\Delta}{\Delta + \gamma}(R_n + G)\frac{\gamma}{\Delta + \gamma} \quad (7.4382)$$

$$(1 + 718 \times 10^{-7}\ V)(e_z^o - e_z)$$

where:

ETP = potential evapotranspiration (mm/d)
$\frac{\Delta}{\Delta + \gamma}$ and $\frac{\gamma}{\Delta + \gamma}$ = mean air temperature and elevation correction factors
R_n = net solar radiation (W/m²)
G = mean daily soil heat flux (W/m²)
V = wind velocity at 2 m above ground (m/s)
e_z^o = mean saturated vapor pressure (mbar), mean value of the P_v measured at the maximum and minimum daily air temperature
e_z = saturated vapor pressure (mbar) at mean dewpoint

Notes: (1) When the net solar radiation is given in langleys per day, the multiplication factor (7.4382) becomes 15.36.
(2) If the vapor pressures are given in pascals, insert a 10^{-2} multiplication factor.

(c) Blaney-Criddle equation

$$u = KF = \Sigma kf;\ \text{monthly}\ u = kf$$

where:

u = consumptive use for growing season (mm)
K = empirical consumptive use coefficient for growing period
F = sum of consumptive use factors, f, for the period
$f = p(0.46\ t + 8.13)$
  where t = mean monthly air temperatures (°C)
  and p = mean monthly percentage of annual daytime hours
k = monthly consumptive use coefficient

(d) Filling of soil water reservoir

   Water applied

$$10\,Qt = ad$$

where

   Q = water flow (m³/h)
   t = number of hours of irrigation
   a = irrigated surface area (ha)
   d = depth of water used would cover irrigated land, in millimeters (mm)

   Time of application

$$t = P_W A_s da / 10^3\,Q$$

where

   t = time of application (h)
   P_W = moisture percentage, dry mass basis
   A_s = apparent relative density of soil
   d = depth of coverage (mm)
   a = irrigated area (ha)
   Q = flow (m³/h)

(e) Water power

$$P = 2.733\,Qh$$

where:

   P = power (W)
   Q = discharge (m³/s)
   h = vertical lift (m)

Irrigation Applications

(a) Submerged orifice

$$Q = 61 \times 10^{-3}\,A\sqrt{2gh}$$

where:

   Q = discharge (L/s)
   A = orifice area (mm²)
   g = gravitational acceleration (m/s²)
   h = water column head (m)

CHAPTER 6—ENGINEERING PROBLEMS AND FORMULAE

(b) Suppressed weir (Francis formula)

$$Q = 184 \, Lh^{1.5}$$

where:

Q = discharge (L/s)
h = total head (m)
L = length of weir (mm)

(c) Trapezoidal weir

$$Q = 186 \, Lh^{1.5}$$

Units defined as per item b.

(d) Triangular weir, 90°

$$Q = 138 \, h^{2.5}$$

Units defined as per item b.

(e) Rectangular weir

A rectangular weir has a crest length of 5.5 m and is operating with a head of 3.6 meters of water. Using the SI version of the rectangular weir formula:

(i) Determine the discharge in cubic dekameters per day
(ii) What will the discharge be in liters per second for a head of 200 mm?

Solution:

(i) $$Q = 158.8 \left(L - \frac{h}{5}\right) h^{1.5}$$

where

Q = discharge, (dam^3/d)
L = crest length, in meters
h = hydraulic head, in meters

$$Q = 158.8 \left(5.5 - \frac{3.6}{5}\right) (3.6)^{1.5}$$
$$= 158.8 \, (4.78)(6.83)$$
$$= 5185 \, \text{dam}^3/\text{d}$$

(ii)

$$200 \text{ mm} = 0.2 \text{ m}$$
$$1 \text{ L/s} = 10^{-3} \text{ m}^3/\text{s}$$
$$= 10^{-6} \text{ dam}^3/\text{s} = 86.4 \times 10^{-3} \text{ dam}^3/\text{d}$$

Coefficient adjustment: $158.8 \div (86.4 \times 10^{-3}) = 1837.9$

$$Q(\text{L/s}) = 1.838 \times 10^3 \left(5.5 - \frac{0.2}{5}\right)(0.2)^{1.5}$$
$$= 1.838 \times 10^3 \, (5.46)(0.0894)$$
$$= 897.2 \text{ L/s}$$

Atmospheric Water

(a) Absolute humidity

$$P_y = 18 \, e/RT$$

where:

P_y = absolute humidity (kmol/m^3)*
e = vapor pressure (Pa)
R = universal gas constant
T = kelvin temperature

Note: When vapor pressure is given in millibars, the formula becomes

$$P_y = 1800 \, e/RT$$

(b) Specific humidity

$$q = \frac{m_v}{m_v + m_a}$$

where:

q = specific humidity
m_v = mass of water vapor (mg)**
m_a = mass of dry air (mg)**

(c) Saturated vapor pressure

$$e^o = \left(\frac{r^o}{0.622 + r^o}\right) P$$

---

* The kmol/m^3 can be related to mg/L.
** Mass values may also be given in grams or kilograms; both units must be the same.

CHAPTER 6—ENGINEERING PROBLEMS AND FORMULAE

where:

e^o = saturated vapor pressure (Pa)
r^o = saturated mixing ratio
P = total pressure (Pa)
Note: The formula remains unchanged when expressing P and e^o in millibars.

(d) Relative humidity (percent value)

$$R.H. = 100\, e/e^o$$

The "e" symbols as previously defined.

Miscellaneous

(a) Salinity concentration

$$T = \frac{PV}{K}$$

where:

T = tons (metric tons) of salt in volume of water V
P = water salinity in parts per million (mg/L)
V = volume of water in acre-feet (dam^3)
K = a constant; it equals 737 for customary units
Note: 1 mg/L $\simeq$ 1 part per million (p/m)
    Translate this equation to determine the value of K which will permit the use of the SI units shown in parenthesis.

$$\text{tons} = \frac{\text{p/m} \times \text{acre-ft}}{737} \quad \text{(Customary)} \quad (1)$$

$$\text{metric tons} = \frac{\text{mg/L} \times \text{dam}^3}{K} \quad \text{(SI metric)} \quad (2)$$

Divide (2) by (1)

$$\frac{\text{metric tons}}{\text{tons}} = \left(\frac{\text{dam}^3}{\text{acre-ft}}\right) \left(\frac{737}{K}\right)$$

$$K = (737) \left(\frac{\text{dam}^3}{\text{acre-ft}}\right) \left(\frac{\text{tons}}{\text{metric tons}}\right)$$

Conversion relationships: dam^3 = acre-ft $\times$ 1.233 48
                         tons = metric tons $\times$ 1.102

Substituting:

$$K = 737 \times 1.2335 \times 1.10$$
$$K = 1000$$
$$T = \frac{PV}{1000} \qquad \text{(SI units)}$$

Alternatively:

V = volume in cubic dekameters (dam^3)
No. of liters = $V \times 10^6$
milligrams of salt = $PV \times 10^6$
metric tons of salt $(T) = \dfrac{PV}{1000}$

(b) Biological oxygen demand
   The biological oxygen demand surface loading on a reservoir is 180 (lb/acre)/d. What is this in (kg/ha)/d?

$$L = \frac{M}{AT}$$

where:

L = loading rate [(lb/acre)/d]
M = mass applied in pounds (mass)
A = area to which mass is applied, acres
T = time of application in days

$$180 = \frac{\text{lb}}{\text{acres} \times \text{day}} \qquad (1)$$

$$L_{SI} = \frac{\text{kg}}{\text{ha} \times \text{d}} \qquad (2)$$

Divide (2) by (1)

$$L_{SI} = 180 \left(\frac{\text{kg}}{\text{lb}}\right)\left(\frac{\text{acres}}{\text{ha}}\right)$$

Conversion relationships: lb = kg × 2.20
acres = ha × 2.47

CHAPTER 6—ENGINEERING PROBLEMS AND FORMULAE

Substituting:

$$L_{SI} = 180 \times \frac{1}{2.20} \times 2.47$$

$$L_{SI} = 202 \text{ (kg/ha)/d}$$

(c) Ideal gas equation of state

$$pv = RT$$

where:

p = pressure of gas, in pascals
v = specific volume (m^3/kg)
R = gas constant [J/(kg·K)]
T = thermodynamic temperature, in kelvins

What is the specific volume of air at a pressure of 185 kPa and a temperature of 35 °C? Assume that air is an ideal gas, therefore:

$$v = RT/p$$
$$T = 35 + 273.15 = 308.15 \text{ K}$$
$$p = 185 \text{ kPa} = 185\ 000 \text{ N/m}^2$$
$$R = 287 \text{ J/(kg·K)} = 287 \text{ N·m/(kg·K)} \text{ (Table data)}$$

$$v = \frac{287 \frac{N \cdot m}{kg \cdot K} \times 308.15 \text{ K}}{185\ 000 \text{ N/m}^2}$$

$$= 0.478 \text{ m}^3/\text{kg}$$

The density of the air = v^{-1} = 2.09 kg/m^3.

BIBLIOGRAPHY

[1] Anton, W. F., "Preliminary Report to AWWA Metrication Committee on Metric Units and Sizes," sixth draft, American Water Works Association, October 1976.

[2] "Architectural and Building Drawings—Presentation of Drawings—Scales," ISO 1047, first edition, International Organization for Standardization, Switzerland, September 1973.

[3] ASME B1 Report, "ISO Metric Screw Threads," American Society of Mechanical Engineers, New York, 1972.

[4] "ASME Orientation and Guide for Use of SI (Metric) Units," ASME SI-1, seventh edition, American Society of Mechanical Engineers, New York, September 1976.

[5] ASME Text Booklet: "SI Units in Dynamics," ASME SI-3, first edition, American Society of Mechanical Engineers, New York, December 1975.

[6] ASME Text Booklet: "SI Units in Fluid Mechanics," ASME SI-5, first edition, American Society of Mechanical Engineers, New York, January 1976.

[7] ASME Text Booklet: "SI Units in Kinematics," ASME SI-6, first edition, American Society of Mechanical Engineers, New York, April 1976.

[8] ASME Text Booklet: "SI Units in Strength of Materials," ASME SI-2, second edition, American Society of Mechanical Engineers, New York, July 1976.

[9] ASME Text Booklet: "SI Units in Thermodynamics," ASME SI-4, first edition, American Society of Mechanical Engineers, New York, November 1976.

[10] ASME Text Booklet: "SI Units in Vibration," ASME SI-8, first edition, American Society of Mechanical Engineers, New York, March 1976.

[11] "Calculations in SI Units," Metrication in the Construction Industry, Ministry of Public Building and Works, London, England, 1970.

[12] "Canadian Metric Practice Guide," CAN3-Z234.1-76, Canadian Standards Association, Rexdale, Ontario, Canada, February 1976.

[13] Dresner, Stephen, "Units of Measurement," Hastings House, New York, 1972.

[14] "Energy Reference Handbook," first edition, Government Institutes, Inc., Washington, D.C., 1974.

[15] "Engineering Drawing Dimensioning," ISO R129, first edition, International Organization for Standardization, Switzerland, September 1959.

[16] Erisman, R. J., "Reducing Metric Conversion Errors," Machine Design, pp. 95-99, June 1976.

[17] "Factors for High-Precision Conversion," NBS Letter Circular 1071, National Bureau of Standards, Washington, D.C., July 1976.

[18] "Inscription of Linear and Angular Tolerances," ISO R406, first edition, International Organization for Standardization, Switzerland, December 1964.

[19] "Style Manual," Government Printing Office, Washington, D.C., 1973.

[20] Mechtly, E. A., "The International System of Units—Physical Constants and Conversion Factors," NASA SP-7012, National Aeronautics and Space Administration, Washington, D.C., 1973.

[21] "Metric Conversion Handbook," State Rivers and Water Supply Commission, Australia, undated.

[22] "Metric Editorial Guide," second edition, American National Metric Council, Washington, D.C., July 1975.

[23] "Metric Information," NBS Special Publication 389, National Bureau of Standards, Washington, D.C., August 1975.

[24] "Metric Units of Measure and Style Guide," USMA Publication 8, U.S. Metric Association, Inc., Boulder, Colo., March 1976.

[25] "Metrication Problems in the Construction Codes and Standards Sector," NBS Technical Note 915, National Bureau of Standards, Washington, D.C., June 1976.

[26] "Official Metric System," J. J. Keller and Associates, Inc., Neenah, Wis., 1975.

[27] Potter, J. H., "Steam Charts: SI (Metric) and U.S. Customary Units," ASME SI-10, American Society of Mechanical Engineers, New York, 1976.

[28] "Precision Measurement and Fundamental Constants," NBS Special Publication 343, National Bureau of Standards, Washington, D.C., August 1971.

BIBLIOGRAPHY

[29] Sellers, R. C., "Dun and Bradstreet's Guide to Metric Transition for Managers," T. Y. Crowell Company, New York, 1975.

[30] "SI Units and Recommendations for the Use of Their Multiples and of Certain Other Units," ISO 1000, International Organization for Standardization, American National Standards Institute, New York, 1973.

[31] "Standard for Commonly Used International System (SI) Units in Bureau of Reclamation Practice—Interim Metric Guide," Bureau of Reclamation, Denver, Colo., September 1976.

[32] "Standard for Metric Practice," ASTM E 380-76^E, American Society for Testing and Materials, Philadelphia, Pa., 1976.

[33] "Technical Drawings—Lettering—Part 1: Currently Used Characters," ISO 3098/1, first edition, International Organization for Standardization, Switzerland, April 1974.

[34] "Technical Drawings: Tolerances of Form and of Position—Part I," ISO R1101, first edition, International Organization for Standardization, Switzerland, July 1969.

[35] "Technical Drawings: Tolerance of Form and of Position—Part III," ISO R1660, first edition, International Organization for Standardization, Switzerland, April 1971.

[36] "Technical Drawings: Tolerances of Form and of Position—Part IV," ISO R1661, first edition, International Organization for Standardization, Switzerland, April 1971.

[37] "The International System of Units (SI)," NBS Special Publication 330, National Bureau of Standards, Washington, D.C., July 1974.

[38] "The Metric Guide," second edition, Council of Ministers of Education, Toronto, Canada, March 1976.

[39] "Units and Systems of Weights and Measures: Their Origin, Development, and Present Status," NBS Letter Circular 1035, National Bureau of Standards, Washington, D.C., June 1975.

APPENDIX

Appendix

USGS STATEMENT ON THE PREPARATION OF METRIC BASE MAPS FOR THE NATIONAL MAPPING PROGRAM

In accord with the national intent to convert to the metric system, the Geological Survey will pursue a policy of proceeding with metrication as rapidly as possible compatible with production goals and objectives and with due consideration for map user needs. All new and completely revised small-scale and intermediate-scale maps will be prepared using the SI (International System of Units). Completely revised standard quadrangle maps formerly prepared in the U.S. customary system will now be prepared in the metric system. New standard quadrangle maps will be prepared in either one or a combination of the systems for the time being, depending on the unique situation in each State. The objective is to ultimately prepare all products of the National Mapping Program in the metric system.

The map elements to be shown in metric units are contours, elevations and distances, bathymetric contours and soundings, and the UTM (Universal Transverse Mercator) rectangular coordinate reference system:

Map scales

The scales for maps prepared in the metric system will be 1:25 000, 1:50 000, 1:100 000, 1:250 000, 1:500 000, and 1:1 000 000. The Puerto Rico series will continue to be prepared at 1:20 000 scale.

Contour intervals

The basic contour intervals for the various map scales will be 1, 2, 5, 10, 20, 50, and 100 meters.

Elevations and distances

Elevations will be shown in meters. Distances will be shown in kilometers.

Bathymetric contours and soundings

Bathymetric contours will be shown in meters at intervals appropriate to map scale. Soundings will be shown in meters.

Universal Transverse Mercator (UTM) Grid

The full-line UTM grid will be shown in meters in accordance with provisions contained in the Statement on Reference Systems, dated August 23, 1974.

Planning for conversion to the metric system of the various map series will be accomplished in accordance with the following guidelines and will be coordinated with the States and other Federal agencies as appropriate:

Complete metrication—All map elements will be shown in the metric system. Included in this category are:

- New 1:25 000-scale, 7.5-minute maps on agreement with the State.
- Remapping, at 1:25 000 scale, of areas presently covered by 1:24 000 scale, 7.5-minute maps.
- New 1:50 000-scale, 15-minute and 1:100 000-scale, 1°30" county and standard-format maps (exception may be granted for ongoing county mapping program if States insist).
- Complete revision of 1:250 000-scale series maps and 1:500 000-scale State base series maps.
- New and complete revision of maps in the National Park Series and other special area maps.
- All future new series national base maps.

Partial metrication—This is the preparation of maps where one or more map elements is in the metric system. Included in this category are:

- New 1:24 000-scale 7.5-minute maps prepared with metric contours and elevations in States that prefer metrication but where it is agreed that 1:24 000-scale maps are needed temporarily to maintain scale continuity.
- Standard and interim revision and reprints of existing 1:24 000-scale maps published at 1:25 000 scale on agreement with the State.

Deferred metrication—Partial or complete metrication will be deferred. Maps in this category include:

- New 7.5-minute 1:24 000-scale maps prepared with foot-unit contours in States that prefer delaying converting to the metric system until complete 1:24 000-scale coverage is available for that State.

APPENDIX

Table A-1.—Listing of accepted non-SI units

Name	Symbol	Name	Symbol
astronomical unit	AU	millibar	mbar
bar	bar	minute	min
curie	Ci	minute of arc	'
day	d	nautical mile	nmi
degree of arc	°	parsec	pc
degree Celsius	°C	rad	rad
electronvolt	eV	rontgen	R
gal	Gal	second of arc	"
hectare	ha	standard atmosphere	atm
hour	h	unified atomic mass unit	u
kilowatt hour	kW·h	volt ampere	V·A
knot	kn	volt ampere reactive	V·A (reactive)
liter	L	year	a
metric ton	t		

Table A-2.—List of units not to be used with SI

Name	Symbol	Name	Symbol
angstrom	Å	metric carat	—
barn	b	metric horsepower	hp_M
calorie	cal	mho	Ω^{-1}
centimeter of water	—	micron	μ
dyne	dyn	millimeter of mercury	—
erg	erg	oersted	Oe
fermi	F	phot	ph
gamma	γ	poise	P
gauss	Gs	stere	st
kilocalorie	kcal	stokes	St
kilogram force	kgf	stilb	sb
kilopond	kp	technical atmosphere	—
lambda	λ	tonne	t
langley	—	torr	Torr
maxwell	Mx	X unit	—

Table A-3.—*Units named after scientists*

Unit	Scientist	Country	Dates
ampere	Andre-Marie Ampere	France	1775-1836
coulomb	Charles Augustin de Coulomb	France	1736-1806
degree Celsius	Anders Celsius	Sweden	1701-1744
farad	Michael Faraday	England	1791-1867
henry	Joseph Henry	United States	1797-1878
hertz	Heinrich Rudolph Hertz	Germany	1857-1894
joule	James Prescott Joule	England	1818-1889
kelvin	William Thomson, Lord Kelvin	England	1824-1907
newton	Sir Isaac Newton	England	1642-1727
ohm	Georg Simon Ohm	Germany	1787-1854
pascal	Blaise Pascal	France	1623-1662
siemens	Karl Wilhelm Siemens	Germany (England)	1823-1883
tesla	Nikola Tesla	Croatia (United States)	1856-1943
volt	Count Alessandro Volta	Italy	1745-1827
watt	James Watt	Scotland	1736-1819
weber	Wilhelm Eduard Weber	Germany	1804-1891

Table A-4.—*Mathematical and physical constants*

Quantity (symbol)	Value
Avogadro constant (N_A)	$6.022\ 169 \times 10^{26}$ kmol^{-1}
Bohr magneton (μ_B)	$9.274\ 096 \times 10^{-24}$ J/T
Bohr radius (a_o)	$5.291\ 771\ 5 \times 10^{-11}$ m
Boltzmann constant (k)	$1.380\ 622 \times 10^{-23}$ J/K
Classical electron radius (r_e)	$2.817\ 939 \times 10^{-15}$ m
Compton wavelength of electron (λ_c)	$2.426\ 309\ 6 \times 10^{-12}$ m
Compton wavelength of proton ($\lambda_{c,p}$)	$1.321\ 440\ 9 \times 10^{-15}$ m
Compton wavelength of neutron ($\lambda_{c,n}$)	$1.319\ 621\ 7 \times 10^{-15}$ m
Electron charge (e)	$1.602\ 191\ 7 \times 10^{-19}$ C
Electron charge to mass ratio (e/m_e)	$1.758\ 802\ 8 \times 10^{11}$ C/kg
Electron magnetic moment (μ_e)	$9.284\ 851 \times 10^{-24}$ J/T
Electron rest mass (m_e)	$9.109\ 558 \times 10^{-31}$ kg
Faraday constant (F)	$9.648\ 670 \times 10^{7}$ C/kmol
Fine structure constant (a)	$7.297\ 351 \times 10^{-3}$
First radiation constant ($2\pi hc^2$)	$3.741\ 844 \times 10^{-16}$ W·m^2
Gas constant (R)	$8.314\ 34 \times 10^{3}$ J/(kmol·K)
Gravitational constant (G)	$6.673\ 2 \times 10^{-11}$ N·m^2/kg^2

APPENDIX

Table A-4.—*Mathematical and physical constants*—Continued

Quantity (symbol)	Value
Gyromagnetic ratio protons in H_2O (γ'_p)	$2.675\ 127 \times 10^8$ rad/(s·T)
Gyromagnetic ratio of protons in H_2O corrected for diamagnetism of H_2O (γ_p)	$2.675\ 196\ 5 \times 10^8$ rad/(s·T)
Magnetic flux quantum (Φ_o)	$2.067\ 853\ 8 \times 10^{15}$ Wb
Natural base (e)	2.718 281 828 459
Neutron rest mass (m_n)	$1.674\ 920 \times 10^{-27}$ kg
Nuclear magneton (μ_n)	$5.050\ 951 \times 10^{-27}$ J/T
Permeability of free space (μ_o)	$4\pi \times 10^{-7}$ H/m
Permittivity of free space (ϵ_o)	$8.854\ 182 \times 10^{-12}$ F/m
Pi (π)	3.141 592 653 589
Planck constant (h)	$6.626\ 196 \times 10^{-34}$ J·s
Proton magnetic moment (μ_p)	$1.410\ 620\ 3 \times 10^{-26}$ J/T
Proton rest mass (m_p)	$1.672\ 614 \times 10^{-27}$ kg
Quantum of circulation ($h/2m_e$)	$3.636\ 947\ 10^{-4}$ J·s/kg
Rydberg constant (R_∞)	$1.097\ 373\ 12 \times 10^7$ m^{-1}
Second radiation constant (hc/k)	$1.438\ 833 \times 10^{-2}$ m·K
Speed of light in vacuum (c)	$2.997\ 956 \times 10^8$ m/s
Stefan-Boltzmann constant (σ)	$5.669\ 61 \times 10^{-8}$ W/(m^2·K^4)
Unified atomic mass unit (u)	$1.660\ 531 \times 10^{-27}$ kg
Volume of ideal gas, standard conditions (V_o)	$2.241\ 36 \times 10^1$ m^3/kmol

Table A-5.—*List of abbreviations and symbols for SI design drawings*

The following abbreviations/symbols are selected typical abbreviations common to the design drawings prepared by the Bureau of Reclamation. They represent a combination of abbreviations extracted from ANSI Y1.1, ISO drafting symbols, and abbreviations from past practices within the Bureau. Certain abbreviations have been modified or deleted to avoid conflicts with ISO practices.

Abbreviations should be used only when their meaning is unquestionably clear; when in doubt, spell out the word.

Refer to table A-6 for symbols to be used for chemical elements.

approximate	approx[1]
asbestos cement	AC
assembly	assy
automatic	auto
automation	automn
back to back	B to B
baseline	BL
battery (electrical)	bat

bearing	brg
bench mark	BM
bill of material	B/M
bolt circle	BC
both sides	BS
bottom face	BF
boundary	bdy
breaker	brkr
bypass	byp
cable	ca
caulked joint	Clk.J
cement asbestos	cem asb
center	ctr
centerline	℄
center to center	C to C
chamfer	△ (see table 4-17)
circuit	ckt
circuit breaker	CB
clear	clr
concrete	conc
conductor (multiple, number indicated) example	4/c
conduit	cnd
construction joint	CJ
contraction joint	Cr.J
control joint	Ct.J
corrugated metal pipe	CMP
culvert	culv
current transformer	CT
depth	↧ (see table 4-17)
diagram	diag
diameter	ϕ (see table 4-17)
dowel	dwl
drain	dr
drawing	Dwg or dwg
each	ea
each face	EF
each way	EW
east	E.
elevation	EL
excavate	exc
expansion joint	Exp.J
face to face	F to F
far side	FS
Federal Stock Number	FSN
flange	flg
gage	ga
gearbox	grbx

APPENDIX

generator	gen
gradient	grad
ground	gnd
guardrail	gdr
heel to heel	H to H
hertz	,Hz
hexagon	hex
hexagonal head	hex hd
high point	HPT
high-water line	HWL
hydraulic	hydr
hydrometer	hydm
inlet	inl
inside diameter	Inside ϕ
installation	instl
intake	intk
interface	intfc
International Pipe Standard	IPS
joint	jt
lateral	latl
latitude	lat.
left hand	LH
left side	LS
length	lg
low point	LP
manhole	MH
manufacture	mfr
manufactured	mfd
mark	mk
maximum	max
metal door	metd
minimum	min
National coarse (thd)	NC
National fine (thd)	NF
National pipe thread	NP
National taper pipe (thd)	NPT
natural ground surface	NGS
near face	NF
near side	NS
nominal	nom.
north	N.
not applicable	NA
number	No.
oil circuit breaker	OCB
oncenter	OC
optional construction joint	OCJ
original ground surface	OGS
outside diameter	Outside ϕ

phase	ph
pipeline	ppln
plate	PL
point of curve	PC
point of intersection	PI
point of tangency	PT
polyvinyl chloride	PVC
precast	prcst
procedure	proc
project	proj
projection	pjtn
quantity	qty
radial	rdl
radius	R (see table 4-17)
record	rcd
reference	ref
reinforce	reinf
reinforced concrete	RC
reinforced concrete culvert pipe	RCCP
reinforced concrete pipe	RCP
reinforced concrete pressure pipe	RCPP
reinforcing steel	RST
required	reqd
reservoir	rsvr
right hand	RH
right-of-way	ROW
road	rd
rubber	rbr
scale	sc
section	SEC
sewer	sew
shaft	sft
shutdown	sht dn
slope	slp
soil pipe	SP
solenoid	sol
south	S.
splice	splc
square section	□
stainless steel	SST
standard	Std.
station	STA
steel	stl
structural	strl
support	sprt
surface	surf
switchboard	swbd
switchgear	swgr

symmetrical	symm
tailwater	TW
tangent	tan
taper	tpr
telemeter	TLM
temperature	temp
terminal	term
terminal board	TB
thermocouple	TC
thermostat	thermo
thickness	I (see table 4-17)
thread	thd
transducer	xdcr
transformer	xfmr
transistor	xstr
transmitter	xmtr
tunnel	tnl
typical	typ
unified coarse thread	UNC
vertical	vert
vertical point intersection	VPI
vertical curve	VC
waterline	WL
water surface	WS
west	W.
working point	WP

[1] The abbreviations are presented in the format to be used when the drawing lettering is done in capital and lowercase letters, which is the case within the Bureau of Reclamation. The ANSI standard calls for all capital letters on drawings, so if following this standard, capitalize all lettering with no change to the abbreviation.

Table A-6.—*Chemical elements and their symbols*[1]

Element	Symbol	Element	Symbol	Element	Symbol
Actinium	Ac	Hafnium	Hf	Promethium	Pm
Aluminum	Al	Helium	He	Protactinium	Pa
Americium	Am	Holmium	Ho	Radium	Ra
Antimony	Sb	Hydrogen	H	Radon	Rn
Argon	Ar	Indium	In	Rhenium	Re
Arsenic	As	Iodine	I	Rhodium	Rh
Astatine	At	Iridium	Ir	Rubidium	Rb
Barium	Ba	Iron	Fe	Ruthenium	Ru
Berkelium	Bk	Krypton	Kr	Samarium	Sm
Beryllium	Be	Lanthanum	La	Scandium	Sc
Bismuth	Bi	Lawrencium	Lr	Selenium	Se
Boron	B	Lead	Pb	Silicon	Si
Bromine	Br	Lithium	Li	Silver	Ag
Cadmium	Cd	Lutetium	Lu	Sodium	Na
Calcium	Ca	Magnesium	Mg	Strontium	Sr
Californium	Cf	Manganese	Mn	Sulfur	S
Carbon	C	Mendelevium	Md	Tantalum	Ta
Cerium	Ce	Mercury	Hg	Technetium	Tc
Cesium	Cs	Molybdenum	Mo	Tellurium	Te
Chlorine	Cl	Neodymium	Nd	Terbium	Tb
Chromium	Cr	Neon	Ne	Thallium	Tl
Cobalt	Co	Neptunium	Np	Thorium	Th
Copper	Cu	Nickel	Ni	Thulium	Tm
Curium	Cm	Niobium	Nb	Tin	Sn
Dysprosium	Dy	Nitrogen	N	Titanium	Ti
Einsteinium	Es	Nobelium	No	Tungsten	W
Erbium	Er	Osmium	Os	Uranium	U
Europium	Eu	Oxygen	O	Vanadium	V
Fermium	Fm	Palladium	Pd	Xenon	Xe
Fluorine	F	Phosphorus	P	Ytterbium	Yb
Francium	Fr	Platinum	Pt	Yttrium	Y
Gadolinium	Gd	Plutonium	Pu	Zinc	Zn
Gallium	Ga	Polonium	Po	Zirconium	Zr
Germanium	Ge	Potassium	K		
Gold	Au	Praseodymium	Pr		

[1] These symbols are used for drawings and text.

APPENDIX

Glossary

absolute zero	—The temperature at which a reversible isothermal process occurs without transfer of heat. This temperature is theoretically 0 K (zero kelvin), or $-273.14\,°C$. Absolute zero is one of two reference points for the Kelvin scale.
accuracy	—The degree of conformity of a measured or calculated value to some recognized standard or specified value.
ampere	—The ampere is the SI base unit for electric current.
ANMC	—American National Metric Council
ANSI	—American National Standards Institute
API	—American Petroleum Institute
ASME	—American Society of Mechanical Engineers
ASTM	—American Society for Testing and Materials
AWG	—American wire gage
base unit	—One of the seven units which form the International System of Units
BIPM	—International Bureau of Weights and Measures
candela	—The candela is the SI unit for luminous intensity
CGPM	—General Conference on Weights and Measures
CIPM	—International Committee for Weights and Measures
coulomb	—The coulomb is the quantity of electricity transported in 1 second by a current of 1 ampere.
conversion	—Finding an equivalent value for a dimension or measurement in another measurement system
customary unit	—A unit from the U.S. system of measurement based upon the inch and pound
derived unit	—SI metric unit formed by dividing, multiplying, or raising to powers the SI base and supplementary units, and/or other derived units
dual dimensioning	—Expressing dimensions or measurements in two systems of measurement.
farad	—The farad is the capacitance of two plates between which there exists a potential of 1 volt, when each plate is equally and oppositely charged by a quantity of electricity equal to 1 coulomb.

Glossary—Continued

General Conference on Weights and Measures	—International body for weights and measures created by the Treaty of the Metre in 1875. The United States is a signatory member.
hard conversion	—Conversion of a product or standard based upon the inch modules to a product or standard based upon metric modules
henry	—The henry is the SI measure for inductance.
IFI	—Industrial Fasteners Institute
International Bureau of Weights and Measures	—International center for scientific metrology whose main function is to establish and maintain international measurement standards. Created by the Treaty of the Metre; executes the decisions of the CGPM and CIPM.
International Committee for Weights and Measures	—Created by the Treaty of the Metre; it prepares and executes the CGPM decisions, and supervises BIPM operations.
International Organization for Standardization	—A nongovernmental body established in 1947 to promote/coordinate international standards.
International System of Units	—The modern metric system established by the 1960 General Conference of Weights and Measures.
ISO	—International Organization for Standardization
joule	—The joule is the SI unit of measure for energy.
kelvin	—The kelvin is the SI unit of thermodynamic temperature; it is the fraction 1/273.16 of the triple point of water.
kilogram	—The kilogram is the SI base unit for mass.
lumen	—The lumen is the luminous flux emitted in a solid angle of 1 steradian by a single point source having the intensity of 1 candela.
meter	—The meter is the SI base unit for length.
metrication	—The process of converting to SI metric system.
Metric Conversion Act	—Public Law 94-168 signed on December 23, 1975, by President Ford. Declares a national policy of coordinating the increasing use of the metric system and established a U.S. Metric Board to coordinate the voluntary conversion.
mole	—The mole is the SI unit for amount of substance.
NBS	—National Bureau of Standards
ohm	—The ohm is the measure of electric resistance.
OMFS	—Optimum Metric Fastener System

APPENDIX

Glossary—Continued

newton	—The newton is the SI unit of force.
preferred numbers	—A logical series or sequence of numbers based upon an arithmetic or geometric progression. Used in hard conversion to minimize the number of product sizes.
prefix	—A term added to an SI unit to form a new unit which is a multiple or submultiple of the basic unit.
radian	—The radian is the SI unit of measure for plane angles.
rationalized numbers	—Similar to preferred numbers; converted values rounded to whole numbers which are multiples of 1, 2, 5, 10, 20, etc.
relative density	—This is the recommended terminology to be used in lieu of specific gravity and specific weight. This terminology is already being used by the Bureau of Reclamation as a measure of soil compactness; see relative mass density.
relative mass density	—Reclamation uses this terminology in lieu of relative density, specific weight, and specific gravity in describing the dimensionless density ratio between a substance and water or air, as required.
rounding	—The processing of reducing the number of significant digits in a number according to rules relating to the required accuracy of the value.
scientific notation	—A mathematical technique where very large or very small numbers are expressed as a number between 1 and 10 times a power of 10.
second	—The second is the SI base unit for time.
SI	—International System of Units
significant digit	—Any digit necessary to define a specific value or quantity to a required precision.
soft conversion	—The changing of all customary measurements to their SI metric equivalents; involves no physical changes.
"specific"	—The term "specific" often refers to physical quantities which are measured in a "per unit mass" value. In SI, this is a "per kilogram" measure; examples include specific volume (m^3/kg) and specific energy (J/kg). Also, used to indicate the ratio of a physical quantity comparing it with the like measure associated with water, for example, specific gravity.

Glossary—Continued

steradian	—The steradian is the SI unit for solid angles.
STP	—Standard temperature and pressure.
supplementary units	—These consist of the radian and the steradian.
symbols	—The short forms of metric units.
unit multiple	—The unit formed by adding a prefix to a SI unit.
triple point of water	—This is the single point combination of pressure and temperature at which pure water exists in three states: solid, liquid, and gas. The TP absolute pressure is 610.615 Pa and the temperature is 273.16 K (0.01 °C). The TP temperature is an arbitrary number used to define the kelvin temperature scale.
U.S. customary unit	—A unit from the system of measurement based upon the inch and pound
U.S. Metric Board	—This 17-member body was created by the Metric Conversion Act to coordinate the voluntary transition to the use of the SI system in the U.S.
volt	—The volt is the SI unit used to measure electric potential.
watt	—The watt is the SI unit used to measure the production/usage rate of all types of energy; the unit for power.
weber	—The weber is the SI measurement unit for magnetic flux.

INDEX

A

Abbreviations
  acceptable, 144
  for design drawings, 261
Absolute scale, (see Thermodynamic scale)
Absolute zero, 267
Absorbed dose, 7
Acceleration
  angular, 45
  conversion factors, 189
  derived unit of, 7
  dynamics, 33, 221
  gravitational, 32
  linear, 45
Accepted non-SI units, 259
Accoustic impedance, 54
Accuracy, 267
Active power, 55
Activity, 10
Adjectival numbers (see Unit modifiers)
Admittance, 55
ADP symbols, 163
Algebraic equations and formulas, 25
Algebraic symbols, 19
American National Metric Council (ANMC), (i), 267
American National Standards Institute (ANSI), (i), (xiii), 267
American Petroleum Institute (API), 37, 267
American Society for Testing and Materials (ASTM), (i), 267
American Society of Mechanical Engineers (ASME), (i), 267
American Wire Gage (AWG), 119, 267
Amount of substance, 2
Ampere, 3, 267
Ampere, Andre-Marie, 14, 260
Angles
  decimal degree, 30
  measurement of, 3, 30
  plane, 2, 5, 45, 213
  solid, 2, 6, 45
  symbol for, 24
  USBR practice, 3
Angstrom, 259
Angular
  acceleration, 7
  derived units, 7
  velocity (see Velocity)
"Are," 23, 45

Area, 29, 45
  common comparisons, 64
  conversion factors, 190
  derived unit, 7
Atmosphere, standard, 175, 214
Atomic number, 26
Atto (see Prefixes)

B

Bar, 51, 214
Bar scales, 137
Barometric pressure, 51, 214
Base units, SI, 1, 267
Beam
  concrete, 229
  deflection, 226
  shear stress, 226
Becquerel, 10
Bending moment, 228
Bernoulli's equation, 234
Bibliography, 251
Bid schedule units, 69
Biological oxygen demand, 248
Blaney-Criddle equation, 243
Block numbers, 16, 167
Boiling point of water, 4

C

Calendar dating, 42
Candela, 3, 267
Capacitance, 55
  electric, 5, 8
Capacity (see Volume)
Capitalization, 13
Celsius, Anders, 14, 42, 260
Celsius scale, 2, 4, 42
Centi (see Prefixes)
Centigrade
  angular measurement, 42
  temperature measurement, 42
Charge, 8
  density, 9, 55
  electric, 8, 56
Chemical
  concentrations, 191
  elements, 26, 266
Classification
  particle, 163
  soil, 156

Clock, 24-hour, 43
Coarse aggregate specification, 89
Coherency, 1, 4
Cohesionless soil pressure, 223
Coined units and symbols, 25
Comma
  as a decimal marker, 16
  omission of, 16
Common fractions, 30
Compound names
  derived units with, 7
Compound units, 20
  "per," 20
  product dot, 21
  slash, 20
Concentration, 7
Concrete construction tolerances, 73
Conductance, 55
  electric, 5, 8
Conductivity, 55
Constants, mathematical and physical, 260
Contours
  Bathymetric, 257
  intervals, 257
Conversion, 267
  factors, 187
  "hard," (xiii)
  numerical values, 175
  procedure, 182
  "soft," (xiii), 269
  tables, 187
Coulomb, 7, 267
Coulomb, Charles Augustin de, 260
Cubic decimeter (see Liter)
Current
  density, 7, 56
  electric, 2
Customary unit, 31, 267

D

Dates, calendar (see Numeric dating)
Deci (see Prefixes)
Decimal marker, 15, 167
  in automatic data processing, 167
  use of comma, 16
  use of period, 15
Decimal series, 184
Decimal system, 1
Definition of base units, 1
  of derived units with special names, 4
  of supplementary units, 4
Degree, 30

Degree Celsius, 13, 42
  subdivisions of, 24
  symbol for, 24, 42
Degree, decimal, 132
Degree, in angle measurement, 24
Degree of curve, 150
Degree, temperature interval, 42
Deka, (see Prefixes)
Density, 49, 54, 192
  derived unit of, 7
  mass, 7, 37
  relative, 37
  relative mass, 37
Derived unit(s), 1, 4, 267
  acceleration, 7
  area, 7
  compound names, 9
  density, 7
  energy, 7
  frequency, 7
  power, 7
  pressure, 7
  special names, 8
  speed, 7
  volume, 7
Design drawings, 125
Diameter, pipe, 61
Digit
  groupings, 16
  in automatic data processing, 167
  significant, 176
Dimensional analysis, 169
Dimension symbols, 144
Dimensionless number, 21
Dimensions, stream, 62
Distances (see Length)
Dot, product
  in compound symbols, 21, 28
  in multiplication, 21
Drafting
  practices, 129
  scales, 137
Drawing
  abbreviations, 144, 261
  notes, 144
Dual dimensioning, 130, 267
DuBoy's formula, 241
Dynamics, 221
Dynamic viscosity, 9

E

Elasticity, 50
Electric
  charge, 56
  charge density, 9

INDEX

current, 2, 56
dipole moment, 56
energy, 56
field strength, 9, 56
flux density, 9, 56
polarization, 56
potential, 6, 56
Electrical
  capacitance, 8
  charge, 8
  conductance, 8
  conductors, 118
  energy, 8
  inductance, 8
  potential, 6, 8
  resistance, 6, 8
Electricity, 55, 193
  quantity of, 7
Electromagnetic radiation, 58, 195
Elevations, 132, 144, 149, 257
Energy, 7, 30, 49
  content of fuels, 197
  conversion factors, 196, 198
  density, 9
  per area time, 198
Engineering problems, 221
Enthalpy, 52
Entropy, 9, 52
Equations, 19, 25
Equidistant features, 133
Equivalent absorption, 54
Exa (see Prefixes)
Excited states, 27

F

Farad, 5, 8, 267
Faraday, Michael, 260
Femto (see Prefixes)
Finishing specification, 83
Flood hydrology, 238
Flow rate, 49, 61, 199
Fluid dynamics, 234
Force, 7, 31, 49, 201
  conversion factors, 201
  moment of, 50
  per length, 201
  problems, 221
  unit of, 7
Force of gravity, 34
Formula conversions, 169
Formulas, 25, 221
Four-digit numbers, 16
Freezing point of water, 4
Frequency, 7, 48, 54, 201
Fuel consumption, 49
Fuels, energy content, 197

G

Gas equation of state, 249
Gasoline consumption (see Fuel consumption)
General Conference on Weights and Measures (CGPM), (xiii), 268
Giga (see Prefixes)
Glossary, 267
Gon, 42
Grade (angle measure), 42
Grain conversions, 202
Gravity, force of, 34
Gravitational acceleration, 32
Gray, 7
Greek symbols, 11
Guidelines
  conversion, 174
  design drawings, 125
  definition, 174
  editorial, 13
Gypsum wallboard specification, 114

H

Halving sequence, 183
Hard conversion, 169, 268
Heat, 203
  capacity, 9, 52
  flow, 52
  flux density, 9
  quantity, 52
  specific, 52
  transfer coefficient, 52
Heavy, light, 33
Hectare, 29
Hecto (see Prefixes)
Henry, 6, 268
Henry, Joseph, 260
Hertz, 7
Hertz, Heinrich Rudolph, 14, 260
Hydraulic
  conductivity, 61, 207
  head, 61
  pressure, 61
Hydrology, 234
  flood, 238

I

Illuminance, 7, 8
Illumination, 208
Impedance, 56
Inductance, 6, 8, 56
Industrial Fasteners Institute, 268
Inertia, 50, 209

Insulated gypsum wallboard system
  specification, 114
Internal energy, 52
International Bureau of Weights and
  Measures (BIPM), 267, 268
International Committee for Weights and
  Measures, 267
International Organization for Stand-
  ardization (ISO), (i), (xiii), 268
International paper sizes, 151
International System of Units (see SI)
Intervals
  temperature, 42
  time, 42
Ionization number, 26
Ionizing radiations, 60
Irradiance, 9
Irrigation applications, 244
Italic symbols (see Quantity symbols)

J

Jensen-Haise equation, 242
Joule, 7, 30, 268
Joule, James Prescott, 260

K

Kelvin, 2, 268
Kelvin scale (see Thermodynamic scale)
Kelvin, William Thompson, Lord, 14, 260
Kilo (see Prefixes)
"Kilo" (erroneous abbreviation), 24
Kilogram, 2, 268
  force, 33

L

Langley, 58, 243
Length, 210
  base unit of, 2
  conversion factors, 210
  in technical work, 126
Light (see Electromagnetic radiation)
Light, heavy, 33
Lightweight, 33
Light year, 210
Linear
  density, 211
  expansion coefficient, 52
  increments, 185
  measure, 34, 45
  references, 63 (see also Length)
Liter, 22, 34
  spelling, 22
  symbol for, 13, 35, 166

Load concentration, 211
Lumen, 7, 268
Luminance, 7
Luminous
  flux, 7
  intensity, 2
Lux, 7

M

Magnetic
  density, 7, 57
  dipole moment, 57
  field strength, 7, 57
  flux, 7, 57
  flux density, 7, 57
  induction, 56
  moment, 57
  polarization, 57
  potential, 57
  vector potential, 57
Magnetism, 55, 193
Magnetization, 57
Magnetomotive force, 57
Manning's formula, 236
Manning's nomograph, 238
Maps
  metric, 257
  USGS metric policy, 257
Mass
  base unit of, 2
  capacity, 192
  conversion factors, 212
  heavy/light, large/small, 33
  number, 26
  problems, 221
  references, 65
  units of, 50
Materials, 118
Mathematical constants, 260
Mechanical impedance, 54
Mechanics, 221
Mega (see Prefixes)
Meteorological pressure, 51
Meter, 1
  definition of, 1, 268
  in technical work, 126
  spelling of, 22
Metric
  Conversion Act of 1975, (i), 22, 268
  conversion factors, 188
  decimal structure of, 1
  dial calipers, 156
  gages, 154
  policy, USBR, (i), (xiii)
  prefixes in, 10
  screw threads, 118
  symbol, 126

INDEX

ton, 35
units, common, 63
  preferred, 44
Metrication
  complete, 258
  deferred, 258
  definition, 268
  partial, 258
Meyer-Peter Muller equation, 242
Micro (see Prefixes)
Micrometer
  in technical work, 126
  measurement device, 154
  unit of measure, 126
Micron (see Obsolete units)
Milli (see Prefixes)
Millibar, 51, 259
Millimeter
  design unit of measure, 129
  of mercury, 215, 259
Minute
  angle, 30, 45, 132, 259
  time, 42, 46, 259
Miscellaneous numbers, 18
Modifier, unit, 18
Modulus of elasticity, 8, 36, 50
Molar energy, 9
Molar entropy, 9
Mole, 3, 268
Molecular physics, 59
Moment, 31, 50
Moment of force, 9 (see also Torque)
Moment of inertia, 50
Momentum, 50
Multiplication dot (see Product dot)
Myer rating, 240

N

Nano (see Prefixes)
National Bureau of Standards (NBS), (i), 268
Neutrons, 27
Newton, 7, 268
Newton meter, 9, 30, 43
Newton, Sir Isaac, 14, 260
Non-SI units, (xiii)
  obsolete, 259
  permitted for use, 259
Nuclear physics, 60
Number groups, 16
Numerals, style for, 16, 23
Numeric dating, 42
Numerical coefficients, 23
Numerical data, 176
Numerical values, 23

O

Obsolete units, 259
Ohm, 6, 268
Ohm, Georg Simon, 260
Optimum Metric Fastener System (OMFS), 121, 268

P

Paper
  International size, 151
Particle classification, 163
Pascal, 7
Pascal, Blaise, 14, 260
Per, 20
Periods
  decimal marker, 15
  punctuation, 15
  substitute for product dots, 15, 21, 164
Permeability
  electrical, 9, 57, 207
  (see also Hydraulic conductivity)
Permeance, 57
Permittivity, 9, 57
Peta (see Prefixes)
Physical
  chemistry, 59
  constants, 260
Pico (see Prefixes)
Pipe dimensions, 61, 124
Plane angle, 2, 213
Plurals, 15
Polyvinyl-chloride waterstops specification, 98
Potential, electric, 6
Pound, 32
Power, 7, 50, 214
  pump, 224
Power, radiant flux, 8
Precipitation, 62
Preferred numbers, 269
Preferred SI units, 45
Prefixes, SI, 10, 14, 35, 269
  automatic data processing, 166
  list of, 10
  preferred, 14, 35
  symbols for, 10, 167
  usage, 23
Pressure, 7, 36, 51, 214
  common examples, 36
  conversion factors, 214
  saturated vapor, 246
  soil, 223
  units of, 7

Prestressed concrete, 231
Product dot
    in compound symbols, 15, 21, 28, 164
    in multiplication, 21
Public Law 94-168 (see Metric Conversion Act of 1975), (i), 22
Publications
    SI in, 68
Pump power, 224
Punctuation, 15
PVC waterstop specification, 98

Q

Quantity, of electricity, 7
Quantity symbols, 11

R

Rad, 7
Radian, 3, 5, 30, 269
Radiance, 7
Radiant intensity, 7, 52
Radiation, electromagnetic, 58
Radioactivity, 8
Rationalized metric values, 183, 269
Reactance, 56
Reference guides, 63
Relative mass density, 37, 269
Reluctance, 57
Renard preferred number series, 184
Reservoir capacity, 47, 62
Resistance, electric, 6, 8
Resistivity, 57
Roman letters
    unit lettering, 11, 130
    unit symbols, 11, 130
Rotational frequency, 7, 48
Rounding, 174, 269
    considerations, 183
    converted values, 179
    maximum/minimum limits, 183
Rubber waterstop specification, 106
"Rule of Reason," (xiii)

S

Salinity concentration, 247
Sand specification, 95
Saturated vapor pressure, 246
Scale
    map, 257
    ratios, 138, 151
Schoklitsch formula, 241
Scientific constants, 260
Scientific notation, 185, 269
Scientists, units named after, 14, 260
Second
    angle, 30, 45, 132
    time, 3, 46, 269
Section modulus, 51
Sedimentation, 241
Shilling mark (see Slash)
SI
    coherency, 1, 4
    editorial format, 13
    establishment, (xiii)
    multiples and submultiples, 10
    structure, 13
    style notes, 27
Siemens, 5
Siemens, Karl Wilhelm, 260
Sieve sizes, standard, 156
Significant digits, 174, 269
SI units
    base, 1, 267
    derived, 1, 4, 267
    named after scientists, 14, 260
    non-SI units, 259
    prefixes, 10
    preferred, 43
    special names, 5
    supplementary, 1, 3, 270
SI symbols, 2, 7, 11, 16
    base units, 2
    derived units, 7
    prefixes, 10
    supplementary units, 2
    symbol modifiers, 25
Slant (see Slash)
Slanted lettering, 130
Slash, 20, 27
Sloping letters (see Quantity symbols)
Soft conversion, 269
Soil
    classification, 156
    pressure, 223
    water reservoir, 244
Solid angle, 2
Solidus (see Slash)
Sound
    intensity, 54
    power, 54
    power level, 54
    pressure, 54
    pressure level, 54
Spacing 18
Specific, 269
    conductance, 62
    energy, 9
    enthalpy, 52
    entropy, 9, 52
    gravity, 37, 269
    heat capacity, 9, 52

INDEX 277

internal energy, 52
latent heat, 52
volume, 7, 249
weight, 269
Specifications, 67
coarse aggregate, 89
concrete construction tolerances, 73
contract, 67
finishes, 83
sand, 95
SI metric, 69
waterstop, 98, 106
Spelling, 22
Square hectometer (see Hectare)
Standard dimension
sequence, 133
Stations, 132, 149
Steradian, 4, 6, 37, 270
Stream dimensions, 62
Stress, 7, 36, 51, 214
Structural steel
shape designations, 134
Structures, 226
Style, SI, 13
capitalization, 13
compound units, 20
for unit and quantity symbols, 11
numerals, 16
punctuation, 15
Submerged orifice, 244
Superscripts (cubes and squares), 19
Supplementary units, 1, 3, 270
Surface tension, 9, 51
Surface textures, 130, 144
Surveying, 149
Symbol modifiers, 25
Symbols (see also Unit names, symbols)
11, 13, 16, 144, 270
design drawings, 261
drafting, 144
for angles and temperature, 24
for liter, 13, 35
lowercased, 14
position tolerance, 136
quantity symbols, 11
Systeme International d'Unites
(see SI)

T

Temperature, 2, 4, 41, 52, 65
common references, 4, 65
conversion, 216
intervals, 42, 52
measurement of (scales), 4
thermodynamic, 2
Tera (see Prefixes)
Tesla, 7

Tesla, Nikola, 260
Thermal
conductivity, 9, 52
flux density, 7
resistance, 52
resistivity, 53
Thermodynamic scale, 2
Third angle projection, 128
Thompson, Wm., 1st Baron Kelvin, 14, 260
Time, 42, 46, 217
base unit of, 2
intervals, 42
numeric dating, 42
period, 54
24-hour clock, 43
Titles, 13
Tolerances
concrete construction, 73
formats, 133
grades, 121
symbols, form and position, 135
Ton
customary, 212
metric, 35, 63, 65, 212
Tonne (see Metric ton)
Torque, 31, 43, 51, 217
Transition curves, 150
Transmissivity, 218
Treaty of the Metre, (i)
Triple point, 3, 42, 270
Twenty-four hour clock, 43

U

United States Metric Board, (i), 270
Unit modifiers, 18
Unit multiple, 270
Unit names
capitalization, 13
spelling, 22
style for, 13
symbols, 2, 7, 11
Units for design drawings, 126
Universal Transverse Mercator, 257
Upright letters (see Unit names,
symbols)
U.S. customary unit, 270
USGS metric mapping statement, 257

V

Velocity
angular, 7, 46
conversion factors, 218
linear, 7, 46

Viscosity, 51, 219
   dynamic, 9, 51
   kinematic, 51
Volt, 6, 270
Volta, Count Alessandro, 260
Volume, 7, 47
   capacity, 220
   conversion factors, 220
   references, 64
   specific, 7,

W

Water
   density of, 63
   freezing and boiling points, 4
   quality, 62
   triple point, 3, 42, 270
   utilization, 242
Watt, 7, 270
Watt hour, 31
Watt, James, 260
Wave number, 7
Weber, 7, 270
Weber, Wilhelm Eduard, 260
"Weight," 31 (*see also* Mass and Force of gravity)
Weight, specific, 269
Weld symbols, 144
Wire sizes, 118
Work (*see* Energy)

Y

Year, 46, 217